Lecture Notes in Civil Engineering

Volume 779

Lecture Notes in Civil Engineering (LNCE) publishes the latest developments in Civil Engineering—quickly, informally and in top quality. Though original research reported in proceedings and post-proceedings represents the core of LNCE, edited volumes of exceptionally high quality and interest may also be considered for publication. Volumes published in LNCE embrace all aspects and subfields of, as well as new challenges in, Civil Engineering. Topics in the series include:

- Construction and Structural Mechanics
- Building Materials
- Concrete, Steel and Timber Structures
- Geotechnical Engineering
- Earthquake Engineering
- Coastal Engineering
- Ocean and Offshore Engineering; Ships and Floating Structures
- Hydraulics, Hydrology and Water Resources Engineering
- Environmental Engineering and Sustainability
- Structural Health and Monitoring
- Surveying and Geographical Information Systems
- Indoor Environments
- Transportation and Traffic
- Risk Analysis
- Safety and Security

To submit a proposal or request further information, please contact the appropriate Springer Editor:

- Pierpaolo Riva at pierpaolo.riva@springer.com (Europe and Americas);
- Swati Meherishi at swati.meherishi@springer.com (Asia—except China, Australia, and New Zealand);
- Wayne Hu at wayne.hu@springer.com (China).

All books in the series now indexed by Scopus and EI Compendex database!

Vikas Chand Sharma · Pankaj Kumar Tyagi ·
George Redhead · Iasef Md Rian · Anurag Varma
Editors

Tech-Enabled Urbanism and Entrepreneurship for Inclusive Cities

Springer

Editors
Vikas Chand Sharma
University Institute of Architecture
Chandigarh University
Mohali, Punjab, India

George Redhead
Norwich Business School
University of East Anglia
Norwich, UK

Anurag Varma
Liberal Arts and Humanities Cluster
Chandigarh University
Mohali, Punjab, India

Pankaj Kumar Tyagi
University Institute of Tourism, Hospitality and Management
Chandigarh University
Mohali, Punjab, India

Iasef Md Rian
Department of Architecture
Xi'an Jiaotong-Liverpool University
Suzhou, China

ISSN 2366-2557 ISSN 2366-2565 (electronic)
Lecture Notes in Civil Engineering
ISBN 978-981-95-3872-0 ISBN 978-981-95-3873-7 (eBook)
https://doi.org/10.1007/978-981-95-3873-7

This Springer imprint is published by the registered company Springer Nature Singapore Pte Ltd.
The registered company address is: 152 Beach Road, #21-01/04 Gateway East, Singapore 189721, Singapore

Preface

The genesis of this book can be traced to a series of converging observations that began during fieldwork across diverse urban landscapes—from the bustling tech hubs to the heritage cities, from Dubai's ambitious smart city initiatives to the grassroots innovations emerging in Kolkata's informal settlements. As we witnessed communities navigating the promises and pitfalls of digital transformation, a critical gap became evident: while smart city discourse dominated academic and policy circles, the lived experiences of urban residents—particularly marginalized communities—remained largely absent from these conversations. The catalyst came during a 2023 workshop in Chandigarh, where urban planners, technologists, and community leaders gathered to discuss the future of cities. As presentations showcased gleaming smart infrastructure projects, a community organizer from a nearby informal settlement posed a simple yet thoughtful question: *These technologies look impressive, but will they actually help my neighbours access clean water, find decent work, or have a voice in decisions that affect their lives?* This moment crystallized our understanding that technological advancement without intentional inclusion could exacerbate rather than resolve urban inequities.

Existing literature on smart cities often treats inclusion as an afterthought rather than a foundational principle. While abundant research examines smart city technologies and their efficiency gains, relatively few works critically examine how these innovations can be designed, implemented, and governed to advance social justice and democratic participation. We observed a troubling disconnect: technology companies and urban planners were creating 'solutions' without meaningful engagement with the communities most affected by urban challenges. This volume has a threefold purpose. First, it reframes the smart city narrative from an efficiency-driven to an equity-centered approach, demonstrating that technological innovation and social inclusion are not competing priorities but mutually reinforcing imperatives. Second, it elevates the voices and experiences of practitioners, entrepreneurs, and community leaders who are creating inclusive technological solutions at the grassroots level. Third, it aspires to provide actionable frameworks to guide policymakers, urban planners, technologists, and social entrepreneurs toward more inclusive approaches to tech-enabled urban development.

Diverse research approaches were employed for this volume, including ethnographic studies in traditional markets, computational analysis of urban data flows, participatory action research with marginalized communities, and policy analysis of smart city initiatives across continents. These diverse perspectives enriched our editorial approach while also creating productive tensions that strengthened the final product. Our numerous editorial meetings—conducted across time zones and cultural contexts—became laboratories for the kind of inclusive dialogue we advocate throughout this volume.

Through 13 chapters organized in three interconnected parts, contributors examine the complex relationship between technological innovation, social inclusivity, and urban development across various cultural and economic contexts.

Digital Foundations and Approaches, establishes conceptual foundations through rigorous exploration of how emerging technologies reshape urban governance, workforce dynamics, and housing accessibility. From Pareek and Sharma's examination of tech-enabled urban empowerment in Rajasthan to Zhang and Choi's analysis of AI-aided urbanism, these chapters reveal how digital platforms and artificial intelligence can enhance service delivery, encourage civic engagement, and reduce structural inequalities when thoughtfully implemented within local governance frameworks.

Entrepreneurship and Social Innovation for Inclusive Urban Futures, delves into the transformative potential of technology-driven social innovation through case studies that demonstrate how urban technologies facilitate participatory governance and sustainable development. Islam and colleagues' comparative analysis of digital citizen hubs across Barcelona, Amsterdam, and Singapore, alongside Baek and Rian's reimagining of traditional wet markets as digitally-enabled urban commons, showcases how artificial intelligence, IoT sensors, and participatory digital platforms can create more inclusive and culturally embedded public spaces.

Contextual Applications, addresses the practical implementation of inclusive tech-enabled urbanism through examinations of heritage revitalization, platform-based food economies, and smart policy frameworks. From Kurnaz's exploration of digital democratization in historic cities to Sharma and Babber's comprehensive framework for smart policies that enhance equitable urban entrepreneurship, these chapters provide concrete methodologies for ensuring that technological advancement benefits all citizens equally.

Throughout this volume, several key themes emerge: the repositioning of technology as a strategic tool for democratizing urban access rather than merely optimizing efficiency; the emergence of new participatory governance models that integrate digital platforms with community engagement; the critical importance of addressing digital divides and ensuring equitable access to technological benefits; and the potential for entrepreneurial approaches to create sustainable solutions that bridge innovation and social justice. This volume does not offer simplistic solutions to complex urban challenges. Instead, it invites readers into ongoing conversations about how technological innovation and entrepreneurial strategies can be harnessed to create cities that are not only technologically advanced but also equitable, sustainable, and human-centered. We recognize that these conversations are necessarily interdisciplinary, requiring the voices of technologists, urban planners, social entrepreneurs, policy makers, and community stakeholders working together toward shared goals of inclusive urban development.

We hope readers will engage with these ideas not as passive consumers but as active contributors to the evolving field of inclusive tech-enabled urbanism. Our gratitude extends to all contributors who shared their insights throughout this collaborative process, as well as to the communities, organizations, and individuals worldwide whose work provides the foundation for this analysis. We believe their continued engagement is essential for translating these ideas into meaningful change. This volume is offered in the hope that it will inspire further research, practice, and policy development at the intersection of technology, entrepreneurship, and inclusive urbanism. May it contribute to the creation of cities that are equitable, sustainable, and reflective of our highest aspirations for human flourishing.

Mohali, India — Vikas Chand Sharma
Mohali, India — Dr. Pankaj Kumar Tyagi
Norwich, UK — Dr. George Redhead
Suzhou, China — Dr. Iasef Md Rian
Mohali, India — Dr. Anurag Varma

Contents

Contributors

Syed Immamul Ansarullah Department of Management Studies, University of Kashmir, North Campus, Delina, Baramulla, Jammu and Kashmir, India

Anisha Arora University Centre for Research and Development, Chandigarh University, Mohali, India

Kirti Babber University Centre of Research and Development, Chandigarh University, Mohali, India

MinJoo Baek Department of Architecture, Xi'an Jiaotong-Liverpool University, Suzhou, China

Sampurna Biswas Department of Architecture and Planning, National Institute of Technology Raipur, Raipur, Chhattisgarh, India

Hee Sun Choi School of Design, The Hong Kong Polytechnic University, Hong Kong, China

Shohreh Ezzatian Art University of Isfahan, Isfahan, Iran

Sumaira Gulzar Department of Management Studies, University of Kashmir, North Campus, Delina, Baramulla, Jammu and Kashmir, India

Marzieh Hadi Tarh_O_Memari Company, Isfahan, Iran

Arash Hajrasooliha Tarh_O_Memari Company, Isfahan, Iran

Sheikh Ikhlaq Cyber- Enterprise Security, Deloitte, London, England

Muhammad Idreesul Islam Department of Management Studies, University of Kashmir, North Campus, Delina, Baramulla, Jammu and Kashmir, India

B. V. Jayanthi School of Management Studies, CBIT, Hyderabad, India

Rajwinder Kaur Apex Institute of Management, Chandigarh University, Mohali, Punjab, India

Ahmet Kurnaz Istanbul Sabahattin Zaim University, Istanbul, Turkey

Iasef Md Rian Department of Architecture, Xi'an Jiaotong-Liverpool University, Suzhou, China

Enayat Allah MohagheghNasab Art University of Isfahan, Isfahan, Iran

Sanjay Nadkarni Plekhanov University, Dubai, United Arab Emirates; Thinkers Garden, Vienna, Austria

B. Neeraja School of Management Studies, CBIT, Hyderabad, India

Khair Ul Nisa Department of Computer Science and Artificial Intelligence, University of Bisha, Asir Region, Bisha, Saudi Arabia

Ritu Pareek Department of Humanities, BIT Mesra, Jaipur Campus, Jaipur, India; Birla Institute of Technology, Mesra, Jaipur Campus, Jaipur, India

Sayon Pramanik Department of Architecture and Planning, National Institute of Technology Raipur, Raipur, Chhattisgarh, India

Mohammad Rahimi Azad University of Isfahan, NajafAbad, Iran

Piyal Roy Brainware University, Kolkata, India

Anju Sharma Department of Computer Science, BIT Mesra, Jaipur Campus, Jaipur, India;
Birla Institute of Technology, Mesra, Jaipur Campus, Jaipur, India

Anmol Sharma Department of Management, Chandigarh School of Business, Chandigarh Group of Colleges, Jhanjeri, Punjab, India

Vikas Chand Sharma University Institute of Architecture, Chandigarh University, Mohali, Punjab, India

Tyba Yousuf Department of Computer Science, SSM College of Engineering, Komarapalayam, Namakkal, India

Wang Zhang School of Design, The Hong Kong Polytechnic University, Hong Kong, China

Digital Foundations and Approaches

From Data to Dignity: Tech-Enabled Urban Empowerment in Rajasthan

Ritu Pareek and Anju Sharma

Abstract Amid growing urbanization and limited resources, Rajasthan stands out as a case study in how digital entrepreneurship and technology-driven urbanism can support inclusive and sustainable development. This chapter explores how technology-enabled interventions—when tailored to local governance frameworks and social needs—can enhance service delivery, encourage civic engagement, and reduce structural inequalities in urban areas. Using a qualitative case study approach, it examines initiatives such as Jodhpur's climate-resilient irrigation programs, Udaipur's AI-based traffic systems, and Jaipur's smart governance platforms. The analysis reveals how digital platforms, artificial intelligence, and public–private collaborations are reshaping urban mobility, access to information, and livelihood opportunities for marginalized populations. By situating these interventions within their sociopolitical context, the chapter proposes a scalable model of inclusive digital urbanism that blends sustainability, responsive governance, and grassroots innovation. The findings contribute to broader conversations on how emerging technologies can be harnessed to foster resilience and equity in rapidly urbanizing regions across the Global South.

Keywords Urbanization · Entrepreneurship · Resilient · Governance · Marginalized

R. Pareek
Department of Humanities, BIT Mesra, Jaipur Campus, Jaipur, India
e-mail: ritupareek@bitmesra.ac.in

A. Sharma (✉)
Department of Computer Science, BIT Mesra, Jaipur Campus, Jaipur, India
e-mail: anjusharma@bitmesra.ac.in

V. Chand Sharma et al. (eds.), *Tech-Enabled Urbanism and Entrepreneurship for Inclusive Cities*, Lecture Notes in Civil Engineering 779,
https://doi.org/10.1007/978-981-95-3873-7_1

1 Introduction

Rapid urbanization and enduring inequality in Indian cities make it imperative to shift from infrastructure-led development to frameworks that prioritize fairness, resilience, and inclusion. Rajasthan provides an essential prism through which to examine how digital technology, state-driven innovation, and grassroots entrepreneurship are transforming urban futures because of its ecological vulnerabilities and social inequities.

The influence of smart city projects in Jaipur, Udaipur, and Jodhpur on urban inclusion is critically examined in this chapter. These programs range from AI-enabled traffic systems to IoT-based irrigation and digital welfare platforms. It questions whether caste, gender, income, and geographic structural injustices are addressed by these measures. The chapter examines the transformative possibilities and constraints of tech-enabled urbanism using the RITE-U Model—Resilience, Inclusion, Technology Integration, Equity, and Urban Sustainability—as an analytical framework. It assesses whether innovations enhance access, participation, and quality of life, especially for marginalized populations, by bridging the gap between macro-policy interventions and local realities. The investigation highlights important issues, such as digital exclusion, linguistic obstacles, and uneven scalability, rather than providing a celebratory narrative. As a test site and possible model for inclusive urban development in the Global South, Rajasthan underscores the need to incorporate equity-focused frameworks, like RITE-U, into planning, policy, and impact analysis.

2 Review of Literature

With calls for sustainable and inclusive development, digital expansion, and rapid urbanization, the discourse on tech-enabled urbanism has intensified. National evaluations like the Digital Governance Index highlight disparities in implementation across states, with Rajasthan emerging as a hub for digital innovation that promotes access and accountability [1]. While scholars have emphasized the importance of digital infrastructure in enabling data-driven governance [2], others caution against top-down technocratic approaches that neglect local sociopolitical dynamics and risk exacerbating digital divides [3].

India's Smart Cities Mission, launched in 2015, was designed to foster citizen-centric governance and sustainable infrastructure [4]. Cities such as Pune and Surat have implemented AI-based surveillance and GIS-enabled planning tools [5]. However, critics argue that in the absence of inclusive policy frameworks, such technologies may inadvertently marginalize vulnerable groups such as women, migrants, and informal workers [6].

In Rajasthan, flagship platforms like the Jan Soochna Portal and initiatives such as iStart have positioned the state at the forefront of digital governance [7]. Despite this, scholars note that measurable improvements in gender equity and urban poverty reduction remain limited [8]. The issue of digital exclusion, especially among women, the elderly, and rural poor, persists—an observation supported by rights-based frameworks advocating for "techno-social equity" [9, 10].

Though Rajasthan has made progress in renewable energy through projects like the Bhadla Solar Park, integration into comprehensive urban systems is still in its nascent stages [11]. Moreover, AI and IoT-based interventions in urban services such as traffic, water, and energy management present new challenges related to equity and data privacy [12]. Collectively, these studies highlight both the transformative potential and limitations of tech-driven urban initiatives. There is a growing scholarly consensus that meaningful change hinges on embedding technology within inclusive, participatory, and context-sensitive governance models.

3 Context: Urban Challenges and Digital Interventions in Rajasthan

Rajasthan, India's largest state by area, is undergoing a significant transformation as its cities rapidly urbanize in response to growing infrastructure, social, and environmental challenges. With an annual urban population growth rate of approximately 2.3%, major urban centers like Jaipur, Udaipur, Jodhpur, and Ajmer face immense pressure on civic infrastructure, employment systems, public service delivery, and water resources. These challenges are compounded by the state's arid geography—over 60% of which is desert—intensifying vulnerabilities to climate shocks, water scarcity, and socio-economic disparities.

In response, Rajasthan has emerged as a key player within the Digital India and Smart Cities Mission frameworks by integrating data-driven planning with entrepreneurial innovation. Cities like Jaipur and Jodhpur have introduced smart city solutions to alleviate infrastructure strain through IoT-based traffic management, e-waste systems, and digital public service platforms. According to progress reports, these cities have demonstrated notable advances in municipal efficiency, urban mobility, and water conservation [13].

As per the India Smart Cities Mission Annual Report 2022–23 [14], cities such as Jaipur and Udaipur have completed over 70% of their planned projects—including adaptive traffic lights and IoT-based waste management systems—positioning Rajasthan among the country's leading performers [15]. Nationwide, more than 7,000 smart city projects have been executed across 100 cities, with Rajasthan cities ranking high in both project tendering and completion rates [16].

Figure 1 illustrates the RITE-U Model concerning Rajasthan's smart urban initiatives. It visually connects each pillar—Resilience, Inclusion, Technology Integration, Equity, and Urban Sustainability—with corresponding smart city interventions analyzed in this chapter.

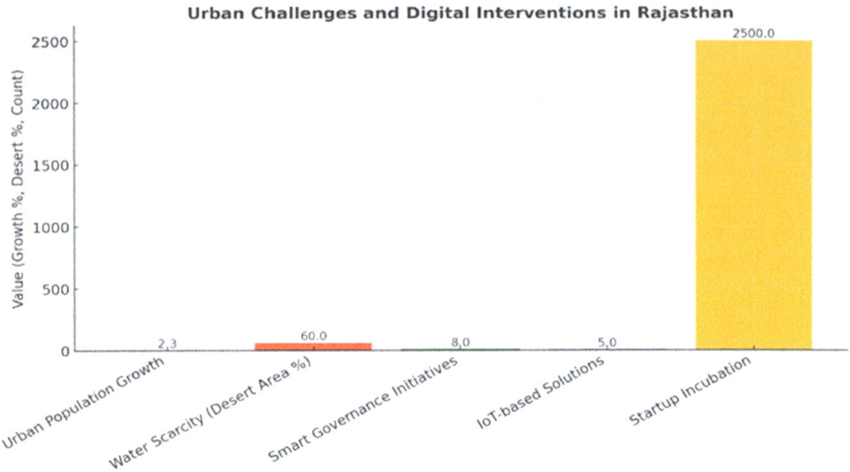

Fig. 1 Urban challenges and digital interventions in Rajasthan. (*Source* Adapted from [13] and [15])

4 Research Objectives

This inquiry is based on the following objectives:

1. To examine how technology-enabled urbanism is addressing Rajasthan's core urban concerns, especially those about transportation, water management, and the provision of public services.
2. To explore how state-sponsored business initiatives support inclusive economic growth and gender empowerment.
3. To investigate how well digital governance systems can promote transparency, civic engagement, and social inclusion.

4.1 Research Design and Methods

Using both primary and secondary sources, this study employs a qualitative research methodology bolstered by case-based analysis and descriptive analytics. To facilitate an interdisciplinary and contextualized understanding of Rajasthan's urban transformation, the research design is organized around a multi-method approach.

4.1.1 The Case Study Method

4. Case Study Approach and Analytical Framework

Three metropolitan centers—**Jaipur, Udaipur, and Jodhpur**, form the basis of an in-depth case study due to their advanced adoption of smart city technologies and digital entrepreneurship ecosystems. The case studies examine:

- The deployment of **IoT, AI, and GIS-based technologies** and their effects on waste management, water conservation, and traffic optimization.
- The role **of digital governance programs** such as the Bhamashah Yojana and the Jan Soochna Portal in improving access, transparency, and service delivery for marginalized communities.

For example, Rajasthan's **Jan Soochna Portal** illustrates how proactive public information systems can significantly enhance administrative transparency and citizen participation. Empirical findings suggest that this platform has improved awareness of entitlements and access to welfare schemes in underserved districts [17].

4.1.2 Data Sources

The study draws from the following data sources:

- **Government Reports**: White papers from Rajasthan's Department of Information Technology, Urban Development Policy briefs, and Smart Cities Mission performance updates.
- **Policy Documents**: Strategic documents such as *Digital India Vision*, *National Urban Innovation Stack*, and *Rajasthan's Startup & Entrepreneurship Policy*.
- **Secondary Literature**: Peer-reviewed journals, think tank publications, and global indices tracking digital equity, urban resilience, and innovation.

4.1.3 Analytical Framework

Drawing on the conceptual frameworks proposed by Townsend [2], Gurumurthy and Chami [9], and recent studies on the tech-governance interface [18], the analysis adopts a **technology–society interface lens.** This approach views digital technology not merely as an infrastructural tool, but as a mechanism for fostering **inclusive, participatory, and equitable urban ecosystems.**

The framework is structured around three evaluative lenses:

- **Social Equity Lens**: To assess access and impact across caste, class, gender, and geographic lines.
- **Sustainability Metrics:** To evaluate environmental improvements, such as reductions in carbon emissions, water use efficiency, and traffic congestion.
- **Entrepreneurial Impact Lens**: To explore outcomes related to innovation diffusion, inclusive employment generation, and localized economic empowerment.

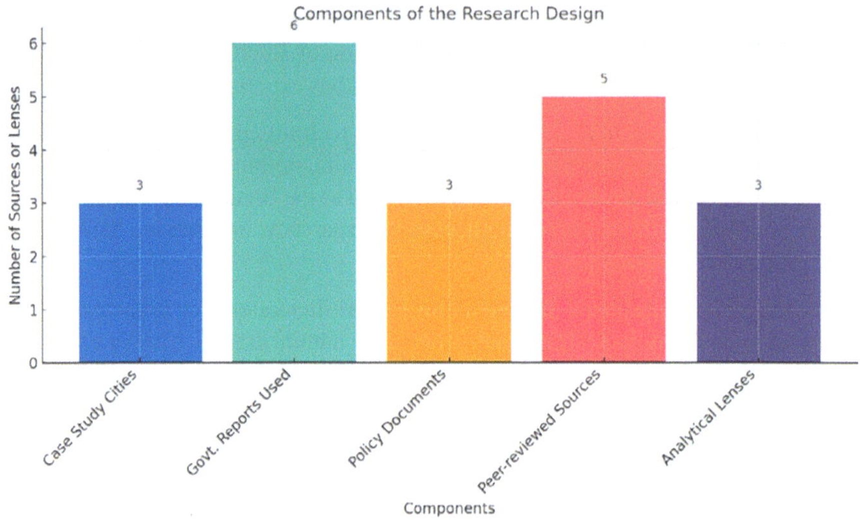

Fig. 2 Urban transformation research design. (*Source* Adapted from [1] and [19])

By employing this multidimensional framework, the chapter moves beyond descriptive documentation and offers a **critical, scalable, and context-sensitive assessment** of Rajasthan's digital urban transformation.

Figure 2 presents a bar chart comparing digital access and exclusion rates across Jaipur, Udaipur, and Jodhpur. As shown in Fig. 2, while Jaipur leads in digital literacy programs, Udaipur shows greater success in e-governance usage, highlighting uneven development patterns across cities.

Urban Population Growth: 2.3% annual increase

Water Scarcity: 60% of the state is deserted

Smart Governance Initiatives: 8 active programs

IoT-based Solutions: 5 key applications

Startup Incubation: 2500+ startups supported

5 Case Studies

Case Study 5.1: Jaipur—Digital Governance and Civic Empowerment:

The capital of Rajasthan, Jaipur, has emerged as a crucial testing ground for smart urban ideas as part of India's Smart Cities Mission. Jaipur is integrating technology, administrative change, and community engagement to create an urban ecosystem that is more inclusive, efficient, and responsive. The GIS-enabled Water Supply Monitoring System and the Integrated Waste Management System are two landmark projects that offer insight into how digital governance may be operationalized to alleviate basic urban service shortages, particularly in underprivileged neighborhoods.

5.1 Case Study: Jaipur—Smart Infrastructure and Inclusive Governance

5.1.1 A Water Supply Monitoring System Powered by GIS: Improving Efficiency and Equity

One of Jaipur's key innovations is the GIS-enabled Water Supply Monitoring System. This system facilitates real-time monitoring of pipeline networks, including the detection of leaks and fluctuations in water pressure. According to the Smart Cities Mission Annual Report [4], this system has significantly improved water supply reliability in informal settlements previously neglected due to aging infrastructure and has reduced non-revenue water losses by 15%. Singh and Dey [20] also confirm these gains in service dependability.

This intervention directly aligns with the **Technology Integration** and **Equity** pillars of the RITE-U model, as it helps bridge service gaps in low-income urban zones. It also contributes to **Urban Sustainability** by enabling predictive maintenance and mitigating water stress in Jaipur's semi-arid environment [21].

5.1.2 Integrated Waste Management System: From Efficiency to Empowerment

Jaipur's Integrated Waste Management System includes GPS-tracked waste collection vehicles and participatory digital tools such as the "Swachhata App," which enables real-time reporting of unattended waste. According to the Urban Development Report [21], this approach has improved collection efficiency by 30% and significantly reduced complaints in densely populated, low-income areas.

A key innovation is the employment of "Swachh Sakhis"—women from disadvantaged communities—who collect household waste door-to-door. This model merges **Inclusion, Equity**, and **Technology Integration** by connecting digital governance to community-based livelihoods, thus addressing structural digital and economic exclusion.

5.1.3 Critical Reflections and Remaining Gaps

Despite these successes, challenges persist. Digital literacy remains low among elderly residents and migrants, with only 38% of households in informal settlements regularly using civic apps, according to a Digital Empowerment Foundation survey [21]. Also, while the "Swachh Sakhis" model fosters inclusion, these workers often remain in precarious, low-wage roles that are not integrated into formal municipal structures. Jaipur's case demonstrates the operationalization of multiple RITE-U pillars through context-aware, participatory urban interventions. However, evolving strategies are essential to tackle persistent gender, digital, and economic inequities.

5.2 Case Study: Udaipur—AI-Driven Urban Mobility and Public Safety

5.2.1 AI-Powered Traffic Management: Efficient and Equitable

Udaipur's transformation includes an AI-powered Traffic Management System using predictive analytics and computer vision. This system has reduced congestion by 22% in key traffic zones [8]. The improvements benefit vulnerable populations such as daily wage workers, emergency responders, and public transit users.

Design elements such as pedestrian-priority crossings and adaptive traffic signals support the elderly and disabled, aligning with **Inclusion** and **Technology Integration.** Furthermore, ambulance response times have decreased by 40%, illustrating technology's life-saving potential in high-density zones.

5.2.2 Civic Data Dashboards: Participation and Responsiveness

Udaipur has implemented civic data dashboards to track and respond to grievances related to infrastructure and utilities. These dashboards have improved service responsiveness by 33% [22]. Reports from iStart Rajasthan [18] suggest that such tools have increased civic participation, particularly among women and youth.

These platforms exemplify **Equity** and **Technology Integration** under the RITE-U framework, showing how digital governance can foster localized accountability and inclusive planning.

5.2.3 Critical Reflections: Where Challenges Persist

Despite efficiency gains, risks remain. Algorithmic bias may result in disproportionate resource allocation favoring affluent zones. Robust oversight mechanisms are necessary to ensure balanced outcomes across all neighborhoods.

Udaipur's innovations reflect the RITE-U principles of **Technology Integration**, **Resilience**, and **Inclusion,** but highlight the need for ongoing monitoring and adaptation to ensure equitable impact.

5.3 *Case Study: Jodhpur—Smart Irrigation and Resilient Infrastructure*

5.3.1 IoT-Based Smart Irrigation: Technology Serving Marginalized Farmers

Jodhpur's smart irrigation initiative uses IoT-enabled soil sensors, automated valves, and weather forecasting to enable precision agriculture. This has improved water-use efficiency by 18% [23]. The initiative notably benefits women-led farming groups and smallholder collectives in peri-urban zones. Improved yields and cost savings have enabled many to transition from subsistence to market-based farming.

This intervention embodies the **Technology Integration**, **Resilience**, and **Equity** pillars of RITE-U, showcasing inclusive agricultural modernization.

5.3.2 Urban Heat Mitigation: Resilience in Informal Settlements

GIS mapping and satellite data have been used to identify heat-vulnerable informal housing zones. Pilot projects involving green roofing, reflective paints, and vertical gardens have reduced indoor temperatures by 3–5 °C [12]. These climate-smart interventions promote **Urban Sustainability** and **Resilience,** improving public health and social protection in underserved areas. However, adoption is hindered by financial barriers, property rights issues, and lack of maintenance.

5.3.3 Critical Reflection: Bridging Ecological Equity and Urban Inclusion

Jodhpur's approach marks a shift from infrastructure-centric to inclusion-focused smart development. Its emphasis on peripheral agriculture and informal housing illustrates how climate-responsive planning can empower historically marginalized communities and fulfill multiple RITE-U objectives (Table 1).

Figure 3 is a chart visualizing the thematic synthesis of case studies using the RITE-U model. As depicted in Fig. 3, Technology Integration and Equity emerged as the most significant pillars influencing inclusive outcomes, while gaps in Resilience and Sustainability signal areas for further intervention [24].

Table 1 Summary of tech-enabled urbanism and entrepreneurship initiatives in Rajasthan

City/initiative	Technology used	Focus area	Key outcomes	Social impact
Jaipur: smart water and waste systems	GIS mapping, IoT, GPS, Mobile apps	Water and waste management	15% reduction in water loss; 30% increase in waste collection efficiency	Improved access in informal settlements; Employment for marginalized women
Udaipur: smart traffic management	AI, Predictive analytics, real-time surveillance	Urban mobility and emergency access	22% reduction in congestion; 40% faster emergency response	Improved mobility for workers; Enhanced pedestrian safety
Jodhpur: smart irrigation and climate resilience	IoT, Satellite imaging, Green roofing	Agriculture & Urban Resilience	18% increase in irrigation efficiency; 3–5 °C reduction in indoor heat levels	Increased yields for smallholders; Better health for urban poor

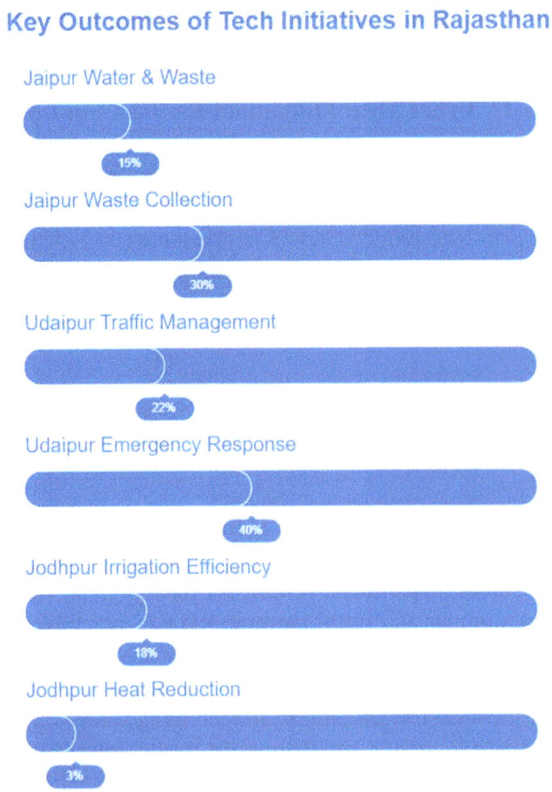

Fig. 3 Key Outcomes of Tech Initiatives in Rajasthan. (*Source* Adapted from [6] and [15])

6 Discussion: Advancing Inclusive Urbanism Through the RITE-U Lens

The evolution of smart and inclusive urbanism in Rajasthan exemplifies how technology, grassroots entrepreneurship, and state-led innovation can collectively transform urban experiences. However, the true measure of such transformation lies not only in infrastructure enhancement but also in dismantling systemic barriers, especially those impacting marginalized populations. This discussion employs the RITE-U Model—Resilience, Inclusion, Technology Integration, Equity, and Urban Sustainability—as a critical evaluative framework to engage with these complexities and bridge gaps identified in existing scholarship.

6.1 Rethinking Governance: From Openness to Self-Determination

Digital governance platforms such as the Jan Soochna Portal are redefining transparency and citizen engagement in Rajasthan. The portal's multilingual, real-time access to over 70 welfare schemes marks a shift from previously opaque, top-down delivery models critiqued for digital exclusion [9]. In tribal districts like Banswara, entitlement awareness rose by 45%, reinforcing how public information access fosters decentralization and trust-building [25].

The Bhamashah Yojana further reflects a targeted push towards digital inclusion for women. However, its impact remains uneven, as highlighted in critiques of technocratic governance models that sideline gender equity [6, 26]. These findings validate such critiques with state-specific empirical evidence, emphasizing the need for gender-sensitive e-governance architecture [27].

6.2 The Digital Imperative of Sustainability

Aligned with the RITE-U Model's resilience pillar, Rajasthan's initiatives like the Bhadla Solar Park and smart irrigation systems demonstrate the potential of climate-tech in semi-arid regions. While prior work has emphasized the park's energy potential [28], this chapter extends the analysis by highlighting its contribution to energy equity in underserved urban peripheries.

Also, the role of AI and IoT in infrastructure—previously seen as tools for operational efficiency [17] is reframed here as instruments of equity when regionally adapted. This analysis calls for integrated assessments linking environmental planning with social justice, a perspective embedded within the RITE-U framework.

6.3 *Entrepreneurship as a Pathway to Inclusion*

Building on previous evaluations of inclusive entrepreneurship [29], this chapter shows how Rajasthan's iStart and Bhamashah Techno Fund programs have incubated over 2,500 startups, with a focus on first-generation and women entrepreneurs [30]. Case studies like Khetify and FinAvish illustrate how low-cost tech-enabled services can address both rural livelihoods and gender gaps in innovation ecosystems.

These initiatives challenge traditional startup paradigms by structuring entrepreneurship to counteract historical exclusions—especially those rooted in caste, gender, and geography [8]. The chapter further introduces the WEP Dashboard's AI-based support for 1.8 lakh women entrepreneurs, addressing a data gap in earlier literature while reinforcing calls for decentralized platform access [10].

6.4 *Recognizing the Digital Divide*

Despite technological advances, digital exclusion remains a structural barrier. A recent study found that 26% of Rajasthan's urban households still lack reliable internet connectivity [12]. Programs like DigiShala, while impactful, remain clustered in Tier 1 cities, leaving peri-urban and rural users underserved.

This section expands on prior critiques [19] by providing concrete geographic illustrations of access disparity and proposing scalable interventions via the RITE-U models such as regional language platforms and decentralized digital hubs—thereby translating conceptual equity into actionable design.

6.5 *RITE-U Model Operationalization: Policy Guidance*

This section operationalizes the RITE-U framework into a strategic policy roadmap—addressing previous criticisms of fragmented implementation in Rajasthan's smart city projects [6]. Key recommendations include digital equity audits, regional innovation hubs, and participatory planning mechanisms.

Evidence from Jaipur and Udaipur suggests that combining civic-tech platforms with socially inclusive planning leads to improved service delivery for marginalized communities. This supports and expands findings on AI-based governance efficacy, contingent on its alignment with equity goals [22].

Ultimately, these findings extend earlier work on smart city implementation by introducing an integrated, intersectional framework that accounts for technological, environmental, and social dimensions. The RITE-U model thus addresses a critical research gap in evaluating smart urbanism through the lens of inclusion and impact in low-resource, high-vulnerability geographies such as Rajasthan.

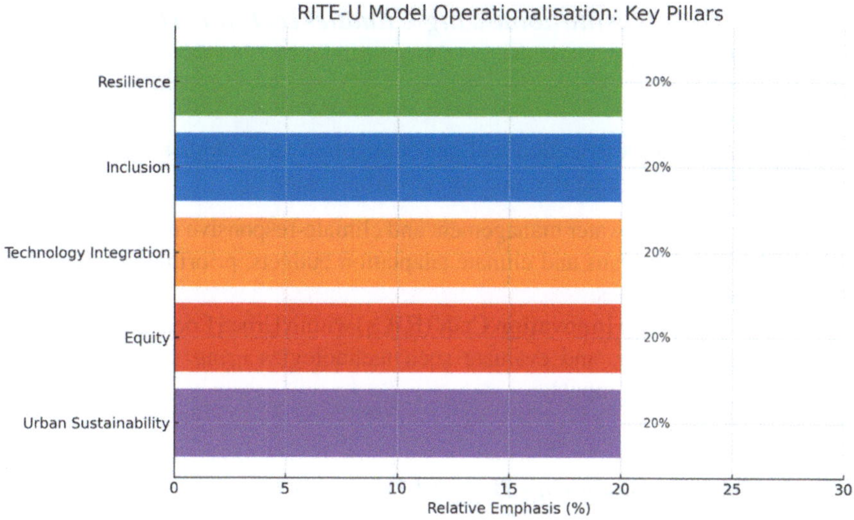

Fig. 4 Enhancing sustainability in arid regions. *Source* By authors

Figure 4 illustrates the proportional emphasis placed on climate-sensitive innovation within Rajasthan's smart urbanist strategies. The allocation of 20% to smart irrigation reflects the significance of government-backed projects in Jodhpur and Udaipur, which use AI and IoT-enabled water management systems. Supported by MNRE (2023) and Smart Cities Dashboard data, these interventions align with the *Resilience* and *Urban Sustainability* pillars of the RITE-U framework, especially within arid and water-stressed geographies.

7 Policy Recommendations: Operationalizing the RITE-U Model for Inclusive Tech-Urbanism

The RITE-U framework—comprising **Resilience, Inclusion, Technology Integration, Equity,** and **Urban Sustainability**—informs the following strategic actions. These recommendations are not presented as universal best practices but are context-specific responses to systemic issues identified in the case studies of Jaipur, Udaipur, and Jodhpur. Each action suggests concrete steps to embed the model within Rajasthan's evolving smart urban ecosystem.

7.1 Resilience: Institutionalizing Climate-Tech for At-Risk Urban Areas

Identified Gap: Smart irrigation and heat mitigation technologies remain at pilot stage and are not integrated into mainstream urban planning.

- Integrate IoT-based water management and climate-responsive infrastructure into municipal master plans and climate adaptation budgets, prioritizing semi-urban and low-income zones.
- Establish **Resilience Innovation Cells (RICs)** within Urban Local Bodies (ULBs) to localize, monitor, and evaluate such technologies using satellite data and community-sourced inputs.

7.2 Inclusion: Designing Beyond Digital Access

Identified Gap: Marginalized groups—including slum dwellers, women, and informal workers—face persistent barriers in accessing and utilizing digital platforms like grievance dashboards and DigiShala.

- Mandate **Digital Equity Audits (DEAs)** across smart city projects to assess user exclusion patterns and root causes.
- Ensure multilingual, voice-based, and offline service options tailored to Tier 2/3 cities and tribal districts.
- Strengthen digital capacity-building programs with targeted outreach in underserved communities.

7.3 Technology Integration: Move from Tools to Systems

Identified Gap: Tech adoption is fragmented; waste, mobility, and water systems operate in silos with limited cross-sector interoperability.

- Establish **Interoperable Urban Data Hubs (IUDHs)** to integrate civic platforms across utilities—water, sanitation, mobility, and health.
- Promote open APIs and standardized civic data protocols to facilitate ecosystem-wide solutions.
- Encourage startup collaboration through innovation programs (e.g., Startup Mahakumbh) that support community-based technology co-creation.

7.4 *Equity: Embedding Caste, Gender, and Occupational Sensitivity*

Identified Gap: Inclusion metrics are vague or absent, especially for vulnerable populations such as the elderly, transgender individuals, and persons with disabilities.

- Institutionalize **Equity Impact Assessments (EIAs)** alongside technology roll-outs to assess differential benefits and exclusions.
- Incentivize equity-oriented innovation through funding challenges hosted on platforms like iStart Rajasthan, encouraging solutions that prioritize structurally disadvantaged groups.

7.5 *Urban Sustainability: Promote Community-Centered Circular Economies*

Identified Gap: Urban sustainability is often narrowly defined by energy or transport metrics, with limited emphasis on socioeconomic resilience or community innovation.

- Expand sustainability frameworks to include circular economy principles, green livelihoods, and participatory environmental initiatives at the ward level.
- Support social enterprises and cooperative models that link environmental goals with inclusive employment generation.

8 Conclusion

The RITE-U Model offers a comprehensive and context-sensitive framework to reimagine smart city development in regions like Rajasthan, where environmental and social vulnerabilities intersect. Unlike traditional infrastructure-heavy models, RITE-U places marginalized communities and ecological sensitivity at the center of tech-enabled urban planning.

Rajasthan's experience—with initiatives like women-led startups, solar energy parks, inclusive e-governance, and regionally adapted smart technologies—demonstrates that systemic transformation is possible when inclusive intent is paired with grassroots innovation. These initiatives exemplify each of RITE-U's five pillars and provide a roadmap for more just and participatory urban governance in the Global South.

However, persistent digital divides across caste, class, gender, and geography necessitate sustained efforts in digital inclusion, local innovation ecosystems, and public–private–community partnerships (PPCPs). Addressing these systemic disparities requires long-term institutional commitment.

As a replicable framework, RITE-U holds potential for guiding smart city initiatives not only within India but across other resource-constrained urban contexts, helping build cities that are not just smart—but also inclusive, equitable, and resilient.

References

1. Digital Empowerment Foundation (2022) Digital inclusion index: measuring India's urban-rural divide in internet access. https://www.defindia.org
2. Townsend AM (2013) Smart cities: big data, civic hackers, and the quest for a New Utopia. W. W. Norton & Company
3. Greenfield A (2013) Against the smart city. Do Projects
4. Ministry of Housing and Urban Affairs (MoHUA) (2022) Smart cities mission annual report. Government of India
5. Bhattacharya S, Sanyal S (2019) Smart cities in India: urban planning and governance in the digital age. Routledge
6. Chatterjee S, Patel V (2020) Inclusion in Indian smart cities: a study of the impact of smart technologies on informal workers. Urban India 40(1):58–74
7. Singh R, Dey A (2021) Transparency through technology: a case study of Rajasthan's Jan Soochna Portal. J Govern Publ Policy 11(2):65–79
8. Sahoo S, Sahu PK (2020) Smart city and urban planning in India: a case study of Jaipur. Urban Dev Sustain 15(2):89–102. https://doi.org/10.1016/j.scs.2020.102311
9. Gurumurthy A, Chami N (2019) Digital technologies and the urban poor: framing a rights-based agenda for inclusive cities. IT for Change
10. UNESCO (2021) Global education monitoring report: technology in education. https://unesdoc.unesco.org
11. Sharma R, Agarwal D (2023) Evaluating startup ecosystems in tier-2 cities: the case of iStart Rajasthan. J Entrepreneur Innov Emerg Econ 9(1):67–81
12. Rathi A, Tiwari A (2022) Techno-social transformations in India's semi-arid cities: examining the smart city interventions in Jodhpur. Sustain Cities Soc 84:103969
13. Government of Rajasthan (2023) Jan Soochna portal dashboard. Department of Information Technology & Communication. https://jansoochna.rajasthan.gov.in
14. Government of Rajasthan (2023) Smart cities progress report. Department of Local Self Government, Jaipur, Udaipur, Jodhpur. https://smartcities.rajasthan.gov.in
15. Ministry of Housing and Urban Affairs (2022) India smart cities mission: annual report 2022–23. Government of India
16. Ministry of Housing and Urban Affairs (2021) Smart cities mission: progress report. Government of India. https://smartcities.gov.in
17. Sharma R (2023) AI and IoT in urban planning: the case of Rajasthan. J Smart Urban 5(1):45–61
18. iStart Rajasthan (2023) iStart annual impact report. Department of IT & Communication, Government of Rajasthan. https://istart.rajasthan.gov.in
19. Mehta A, Srivastava T (2023) Bridging the digital divide: Rajasthan's DigiShala model. Dig India Rev 11(2):24–33
20. Singh M, Dey A (2021) Public service delivery through digital platforms: lessons from Rajasthan. Indian J Publ Admin 67(3):339–353. https://doi.org/10.1177/00195561211039397
21. Government of Rajasthan (2023) Jaipur smart city progress report. Jaipur Smart City Limited
22. Singh A, Bansal D (2023) AI and governance in Indian cities: bridging data, equity, and public service. J Urban Technol 30(1):18–37
23. Digital India Corporation (2023) Digital India: a framework for transformative governance. Ministry of Electronics & IT
24. Udaipur Smart City Limited (2023) Smart traffic management dashboard data. https://udaipursmartcity.in

25. Jain P, Goyal M (2022) Rajasthan Jan Soochna Portal: a model for digital transparency and accountability in public governance. Indian J E-Govern 18(2):45–53
26. Suri V, Mehta A (2023) Urban technology and the gender divide: the role of state-backed platforms in India. Indian J Urban Aff 8(1):34–51
27. Gurumurthy A, Chami N (2019) Towards a feminist social contract: a gendered framework for digital justice. IT for Change. https://itforchange.net
28. Singh R, Mehta K (2021) Solar cities and green infrastructure: lessons from Bhadla Solar Park, Rajasthan. J Sustain Urban Plan 6(3):89–104
29. Mehta R, Kapoor S (2021) Leveraging startups for inclusive growth: a study of government-backed entrepreneurship in Rajasthan. Indian J Publ Admin 67(3):354–372. https://doi.org/10.1177/00195561211040102
30. Press Information Bureau (PIB) (2023) bhamashah techno fund empowers 2,500+ startups. Government of India

AI-Aided Urbanism

Wang Zhang and Hee Sun Choi

Abstract This chapter examines the profound transformation of contemporary urbanism through Artificial Intelligence (AI), reframing its roles and values through a retrospective lens grounded in socio-technical history in the framework of urban development. By comparing the emergence of AI with prior waves of technological innovation since the modern era, this chapter critically examines how heuristic AI systems are redefining the discipline of urban design. The discussion underscores how AI cultivates a deeper, evidence-based comprehension of urban complexity, catalyzing inclusive urban renewal. By investigating AI's foundational influence, the chapter positions AI as an ontological presence embedded across layers of urbanism Through systematic review along with case studies, three pivotal shifts in urban design model are identified: (1) From evidence-informed problem framing to human communication interaction (HCI); (2) data-informed urban design model shaped through collaboration with an expanded network of stakeholders; and (3) new typologies of urban form and transformation due to the materials and design focus shift. Collectively, these transformations contribute to the advancement of a more inclusive, adaptive, and intelligence-driven paradigm of AI-aided urbanism. Finally, by critically evaluating ethical challenges and data justice issues inherent in AI-aided urbanism, the chapter examines the risks and opportunities of AI as a collaborative agent in fostering more equitable, resilient, and opportunity-driven urban futures.

Keywords AI urbanism · Urban design · Socio-technology · Inclusive urbanism · Heuristic AI

W. Zhang (✉) · H. S. Choi
School of Design, The Hong Kong Polytechnic University, Hong Kong, China
e-mail: wang-z.zhang@connect.polyu.hk

V. Chand Sharma et al. (eds.), *Tech-Enabled Urbanism and Entrepreneurship for Inclusive Cities*, Lecture Notes in Civil Engineering 779,
https://doi.org/10.1007/978-981-95-3873-7_2

1 Introduction

The evolution of cities has been marked by the integration of infrastructures with information and communication technologies (ICTs), transitioning from "analog" to "digital" [1]. Enabled by digital technologies such as Digital Twins and Building Information Modeling (BIM), cities have acquired a distinct "smart" identity, setting them apart from the "modern," "metro," and "skyscraper" eras after industrial revolution. By synthesizing 3D models enriched with metadata, including material specifications, structural performance, costs, energy consumption, lifecycle data—and real-time digital replicas of physical urban systems, cities can now integrate static models (e.g., BIM) with dynamic data streams from sensors, satellites, mobility networks, and citizen feedback. Urban challenges like traffic congestion, flooding, energy use, and air quality can thus be simulated, predicted, and managed more effectively. In this data-rich, simulation-driven urbanism, planning, design, and governance are no longer guided solely by static drawings or projections, rather the mass urban data in different fields form complex information networks to understand the cities. This increasing complex urban systems and needs of evaluating the ever-changing urban environments require more dynamic and evolving urban design models that adapt to the city's growth and transformation.

Considering that, since 2020 onwards, a new paradigm has emerged that integrates AI techniques throughout the entire urban lifecycle—from sensing and modeling to planning, design, real-time management, and citizen engagement. A 2022 UN-Habitat report [2] highlights diverse AI applications in urban development and management, ranging from chatbots for public service delivery (24%) to advanced integrations for city governance (35%) and future planning (26%). AI has demonstrated significant benefits, including time savings, improved urban management, enhanced service efficiency, and increased security, addressing challenges in mobility, energy consumption, water and waste management, urban planning, safety, healthcare, and [3]. These intersections with AI are reshaping spatial computing and urban science, empowering researchers and designers to tackle complex urban issues through data-driven learning systems.

While existing literature explores AI's implications in urban research by categorizing its applications into problem-solving, design optimization, design generation, and simulation [4, 5], AI-aided urbanism introduces a paradigm that leverages AI techniques—such as machine learning, optimization algorithms, knowledge reasoning, and real-time embodiment algorithms—to sense, understand, design, and manage cities as dynamic systems. Rather than viewing AI as a back-office optimization tool, this approach positions it as a collaborative partner that expands how planners, designers, and residents frame problems and test solutions across scales, from regional to street level. It elevates socio-spatial phenomena and urban lifestyles in smart cities to higher levels of intelligence.

Although AI applications in urban design and planning have been evaluated for their benefits and limitations through experimental scenarios, pilot projects, and smart city initiatives, a critical gap remains in understanding their fundamental values and meanings within the broader context of urbanism. While some researchers have explored distinctions within AI urbanism, introducing concepts such as autonomy, character, and accountability [6], these studies often focus primarily on definitional distinctions and functional capacities. Other research has highlighted the AI's impacts in intangible areas- such as changes in governance, performance improvements, and shifts in research directions— but they tend to overlook its influence on the physical urban form and the evolving roles of stakeholders directly involved in shaping the built environment [7, 8].

Without a comprehensive analysis of AI's integration into the structural layers of urban development, current literature falls short in offering a holistic understanding of AI-aided urbanism—one that addresses not only the involvement of designers, planners, and policymakers, but also includes the perspectives of an expanded range of actors and disciplines. In this context, it is essential to understand AI's foundational role by examining its interactions across multiple layers of urban development—from its influence on urban design models to its transformation in spatial layouts and urban forms within urban morphologies (Fig. 1).

Building on this critical perspective, this chapter aims to develop a comprehensive understanding of how AI-driven technologies are not just tools, but agents of a paradigm shift in urbanism. It argues that AI is transforming urban design into a more strategic, evidence-based, and human-centric discipline—one that fosters new types of urban forms and unlocks new design models for urban development.

To structure this exploration, the chapter is organized around four core objectives:

1. **Contextual Foundations**: Trace the transformation effects of technological evolvement on urban design models and urban forms, highlighting the pivotal role of AI in redefining contemporary urbanism.
2. **Theoretical Framework**: Recast the roles and values of AI applications in urban design as a heuristic inquiry through complex urban systems in both academic research and practical applications, discussing how it evolved and became unique from traditional digital tools.
3. **Practice Transformation**: Combining case studies in both urban research and practices, the chapter explains the reasons and the ways AI is reshaping urban design model across three dimensions: (i) urban design integrations with human computer interaction (HCI), (ii) transformation of data-informed design model in urban design process (iii) prospective vision of urban morphologies.
4. **Fundamental Challenges**: Critically examine the risks associated with AI urbanism, including ethical concerns and data justice, emphasizing the need for responsible and inclusive frameworks.

By systematic engaging with AI's roles and values in urbanism, this chapter seeks to bridge the gap between technological innovation and profound impacts in urban development, offering a roadmap for creating more inclusive, human-centered, and opportunity-rich urban futures.

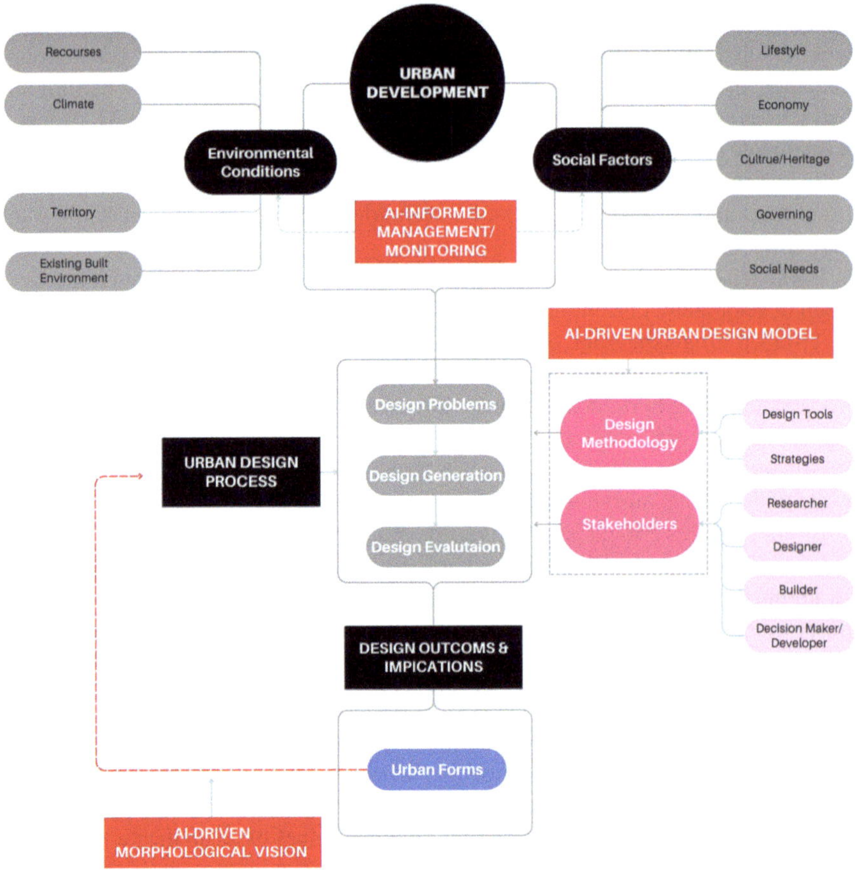

Fig. 1 Framework for understanding urban development in AI-aided urbanism. (*Source* By Authors)

2 Retrospect: Roles and Values of AI in Urbanisms

To elucidate the distinctive role of AI in conceptualizing and implementing contemporary urban development, it is necessary first to examine the socio-technology context driving cities into the AI era, as well as the rationale behind the inevitable integration of AI within urban development. Notably, technological advancements and urban design models continually co-evolve. By comparing AI's transformative potential with previous technological shifts—such as industrialization, computer-aided design (CAD), and smart city digital technologies—this section explores a socio-technical framework: (1) technical innovations → (2) shifts in design methodologies → (3) changes in stakeholder roles → (4) resulting urban forms, providing deeper insights into AI-aided urbanism's unique significance (Table 1).

Table 1 Comparison of the impacts of the four technology waves in urban/architecture design

Technology wave	Core technical improvement	Design methodologies	Expanded stakeholder constellation	Urban forms and layouts
Industrial Revolution	• Mass-produced steel, glass, reinforced concrete	• Engineering-driven rationalism	Industrialists/fabricators, municipal reformers, engineers, architects, financiers	Modernism style buildings, gridiron street grids, skyscraper typology, company towns, separation of land uses
(≈ 1850–1940)	• Mechanized construction and transport	• Early modernism and functionalism		
		• Zoning-based planning		
Computer-Aided Design and BIM	• Digital drafting and 3-D modeling	• Parametric/algorithmic design	Software vendors, contractors, project-management consultants	Non-standard geometries, hi-tech/deconstructivism façades, mass-customized components, mix-used complex
(≈ 1970–2010)	• Parametric accuracy; database-rich BIM	• Performance-based optimization		
		• Digital fabrication workflows		
Smart-City and Big-Data Era	• IoT sensing and real-time urban data streams	• Evidence-based design	Tech firms, platform providers, infrastructure operators, researchers	No significant changes
(≈ 2010–2020)	• Cloud analytics and dashboards	• Scenario simulation and digital twins		
		• Responsive infrastructure management		
AI-Aided Urbanism	Section 2.1 and 2.2	Section 3.1	Section 3.2	Section 3.3
(≈ 2020–present)				

Source By Authors

Although each wave of technological development from the late nineteenth to early twenty-first century has contributed transformative advancements across various domains, their influence on urbanism has varied significantly in scope and depth. Within a socio-technical framework, industrialization and AI stand out for their structural impact—both in reconfiguring stakeholder constellations and generating novel urban forms. This distinguishes them from later technological tools such as Computer-Aided Design (CAD), Building Information Modeling (BIM), Virtual/Augmented Reality (VR/AR), Aerial Photogrammetry, the Internet of Things (IoT), and Digital Twins, which primarily enhance specific aspects of the design and planning process without fundamentally altering the urban paradigm.

Specifically, the industrial revolution revolutionized the use and production of materials. During that period, materials once difficult to work with—such as granite, porphyry, and ivory—became as easily transformable as Malayan rubber, significantly expanding construction possibilities [9]. Similarly, advances in steel, reinforced concrete, and glass manufacturing during the late nineteenth century provided engineers with materials of unprecedented strength and versatility. **However, while these innovations were essential, they were not sufficient on their own to instigate a radical paradigm shift in urbanism.** Initially, these technologies were applied predominantly in engineering contexts—such as bridging vast rivers, roofing large railway stations, and constructing high-rise buildings—rather than transforming architectural aesthetics or urban form [10]. Prestigious buildings in Europe and North America continued to adopt historical revival styles well into the early twentieth century, underscoring a delayed integration of technological potential into architectural expression and urban transformation.

Modern urbanism emerged nearly half a century after the onset of industrialization, when modernist ideology began to integrate engineers and industrial fabricators as key stakeholders in the design process. The modernist principle of "form follows function" emphasized a direct correlation between a building's utilitarian purpose and its form, rejecting excessive ornamentation and reshaping architectural methodologies. This shift catalyzed a deeper interdisciplinary collaboration, wherein architects synthesized aesthetic vision with engineering knowledge and industrial capabilities. In light of that, this integration bridged the gap between technological innovation and aesthetics, ushering in a new unity of form, function, and materiality. As a result, industrialization profoundly influenced what is now recognized as "modern urbanism."

In contrast, subsequent technological developments like computer aided design (CAD) technologies digital tools associated with smart cities in recent decades introduced incremental rather than radical changes in urban design practices. CAD and BIM significantly enhanced drafting precision, geometric complexity, and coordination efficiency, enabling the mass production of residential buildings and the streamlined construction of complex megaprojects [11]. Although CAD/BIM workflows brought in new specialists—software vendors, modeling consultants, and BIM coordinators—their role is primarily technical and downstream. They usually enter the project after the core spatial concept is set and rarely participate in early-stage decision-making. As a result, creative authorship and strategic choices (e.g., massing, program distribution, public-realm intent) remain concentrated in the architect's studio, while the new actors function as service providers who execute and refine pre-established design intent rather than shape it.

With the advent of smart city initiatives, digital technologies like Digital Twins and the Internet of Things (IoT) scaled digitalization from the building to the urban level, embedding big data analytics into all phases of the urban lifecycle. Coupled with CAD/BIM-derived 3D models, VR/AR technologies enabled immersive design experiences, facilitating stakeholder collaboration. Additionally, drones democratized high-precision site analysis, and generative design unlocked unprecedented efficiency gains. Despite these advancements, such technologies primarily enhanced efficiency and problem-solving capacities without fundamentally transforming urban or architectural typologies.

Comparing that, AI technologies, akin to industrialization, offers substantial potential to structurally reconfigure urban design models. To understand this, subsequent sections further explore this potential through the socio-technical framework, drawing on case studies in design research and professional practice to investigate AI's role in reshaping the urban design landscape.

2.1 Technology Improvement: Form Digital to AI-Aided Urbanism

Urban design in the early twenty-first century is characterized by its pursuit of multiple, often interrelated objectives, including economic development, social well-being, and the preservation of cultural identity. These evolving and increasingly complex urban systems necessitate a broad spectrum of design tools and methodologies that support evidence-based, sustainable, and performance-oriented approaches. This shift reflects the growing recognition of the diverse challenges and opportunities inherent in contemporary urban development.

Smart city technologies have emerged as a key response to this complexity, offering integrated and data-driven solutions for shaping and regenerating urban form and function. These technologies enable designers and planners to address a wide range of objectives simultaneously within the design process, thereby improving responsiveness, inclusivity, and efficiency in urban development.

Through the development and applications of digital tools, built environments are increasingly mediated through interconnected layers of information, forming complex data networks. Table 2 illustrates how digital tools tackle complex urban challenges by employing comprehensive data collection methods—capturing, tracing, monitoring, and multilayer storage. For example, GIS techniques such as spatial overlays and hotspot clustering enable analyses of heritage assets and urban morphology, thereby enhancing assessments of local identity [12]. Furthermore, GIS-based environmental evaluations apply raster-weighted overlays to examine pollution, resource flows, and climate risks, integrating interpolation methods and vulnerability indices [13]. Analyses like service-gap and void detection also identify underserved populations, such as seniors, supporting more equitable service provision [14]. Overall, these digital tools empower urban designers to gain a comprehensive understanding of spatial networks, urban systems, and user needs—ultimately addressing environmental sustainability and social integration through more streamlined infrastructure management, transportation optimization, energy efficiency, public safety enhancements, and smart-city development.

Table 2 Digital tools addressing different types of urban design challenges

Challenge type	Digital tools	Role	Sources
Decision-making conflicts	MCDA dashboards	Multi-criteria visualization and decision support	[16]
	Blockchain-style approval ledgers	Transparent process tracking	[12]
Balancing exchange versus use value	3D scanning and photogrammetry	Precise contextual capture	[17]
	AR/VR heritage overlays	Immersive visualization and review	[18]
	Parametric modeling tools	Generative, context-sensitive design	[19]
Context and identity sensitivity	Participatory GIS and crowd-sourced mapping apps	Community-driven spatial data collection	[12]
	Mobile network data analytics	Movement-pattern analysis	[20]
	Digital twins of informal areas	Scenario testing and planning	[21]
Environmental sustainability	IoT sensor networks	Real-time environmental monitoring	[22]
	Urban-scale digital twins and climate-model integration	Climate impact simulation	[21]
	GIS-based risk and vulnerability mapping	Spatial vulnerability assessment	[13]
Social integration and participation	Web-enabled E-forums and chatbots	Inclusive digital engagement	[23]
	Data-driven service-gap analysis tools	Mapping underserved areas	[14]
	Mobile survey apps	In situ community feedback	[24]
Security and resilience	Real-time CCTV	Automated monitoring and anomaly detection	[25]
	Emergency-response simulation platforms	Evacuation modeling and preparedness planning	[26]

Source By Authors

However, prior to the integration of AI, these digital tools used in smart cities and big data analytics were largely confined to quantifying urban phenomena, providing numerical insights without the capacity to generate explanatory narratives or contextual accounts [6]. Users were required to manually interpret vast datasets to extract meaningful insights and contextual value. For instance, early applications of digital twins could provide environmental impact assessments by integrating real-time air quality monitoring, building energy consumption metrics, and microclimate simulations. While the synthesis of real-time data, 3D modeling, and simulation tools enabled more accurate assessments and recommendations, these systems lacked intelligent reasoning capabilities [15]. They were unable to autonomously modify variables or balance complex design objectives—such as aesthetics, economic efficiency, and sustainability—without expert input.

AI technologies, however, significantly enhance the analytical power of these digital methods. By incorporating advanced reasoning and intelligent data processing, AI-enabled tools can autonomously evaluate, simulate, and generate urban environments, morphological configurations, and usage patterns. This marks a fundamental transformation in urban design methodology.

2.2 Heuristic AI and Urbanism

The unique strength of AI lies in its heuristic inquiry, defined by Najafi et al. [27] as an exploratory and intuitive process for navigating complex and uncertain contexts, employing inductive reasoning and analogical thinking.

In parallel, urban design inherently involves heuristic inquiry. For instance, redesigning a city park to enhance usability and enjoyment involves iterative observation, data collection, intuitive ideation, and practical solution formulation. Designers identify issues such as inadequate lighting or insufficient shaded areas through direct site observation. They then utilize reflective methods like sketches, maps, or mood boards to intuitively develop ideas—such as shaded seating, enhanced lighting, or community gardens. These insights are translated into detailed plans, including blueprints, 3D renderings, and community engagement strategies, integrating empirical data, human behavioral insights, and professional judgment.

Importantly, Heuristic AI technologies differ from traditional algorithmic methods, which follow fixed rules, by leveraging rules of thumb and informed guesses [28]. Although no single AI approach universally addresses all heuristic requirements, various AI paradigms provide specific heuristic capabilities in urban design, as classified by Choi & Zhang [4] (Table 3). For instant, Machine Learning (ML) utilizes data-driven heuristics for pattern recognition and design forms generation, and Search and Optimization AI applies evolutionary heuristics to optimize urban layouts. Furthermore, knowledge-based AI incorporates integrated reasoning and expert insights to enable well-informed decision-making. Embodied AI integrates perceptual heuristics through sensorimotor interactions, enabling dynamic human–environment simulations. Collectively, these heuristic approaches enhance adaptive and innovative urban problem-solving strategies. AI-aided urbanism incorporates advanced heuristic approaches into problem-solving and autonomous decision-making process [6, 29].

Notably, AI heuristic methods systematically address urban challenges across three fundamental processes: Problem Formulation (PF), Design Generation (DG), and Design Evaluation (DE). In PF, Convolutional Neural Networks (CNNs) enhance precise spatial analysis, predicting street safety by analyzing correlations between urban features and crime rates, and classifying greenery via satellite imagery. In DG, Artificial Neural Networks (ANNs) and Generative Adversarial Networks (GANs) foster creative urban layouts and visualizations, translating sketches into detailed urban massing plans and street façade designs. During DE, embodied AI such as Agent-Based Modeling (ABM) and Cellular Automata (CA) facilitate rigorous evaluation and iterative refinement, assessing pedestrian accessibility, environmental impacts, and traffic flow simulations.

Figure 2 illustrates practical applications of CNN, GAN, Genetic Algorithm (GA), and ABM models across the three stages of the urban design process. By integrating source data with AI tools and optimizing key parameters, these AI-driven approaches significantly enhance urban outcomes—advancing urban design toward a more automated, intelligent, and reasoned process.

3 AI's Impact on Urban Design: Modeling and Vision

In the preceding sections, we examined the core advancements of AI applications in tackling complex urban design challenges, particularly in comparison to earlier technological developments of the past century. While these findings underscore AI's integration into the urban design process as an augmented tool for heuristic intelligence, it is crucial to recognize that the impact of AI extends far beyond mere technological enhancement in problem-solving. The evolution of design tools represents only the initial threshold of broader transformations in urban design models and urban form. This section delves deeper into these broader impacts, focusing on how AI is reshaping the design discipline, redefining the roles of stakeholders, and addressing the morphological vision of urban development.

Table 3 AI Applications and heuristics methods in urban design

AI Applications in urban design			
AI algorithms	Heuristic methods	Design process	Scope in urban design
Machine learning (ML)	Data-driven learning and pattern recognition		
Artificial neural network (ANN)	Learning, imitation, blending with existing data patterns	DG	Zone scale massing generation
Convolutional neural network (CNN)	Pattern recognition, spatial reasoning	PF	Street safety prediction, green space classification, Street noise classification
Generative adversarial network (GAN)	Pattern generation, creativity, adversarial learning	DG	Street façade generation, masterplan rendering, Street network prediction and generation, Block massing generation, Block buildings layout, Plot building layout generation, Plots typologies prediction
Variational autoencoders (VAE)	Pattern generation, data blending, probabilistic reasoning	DG	Street network generation
Deep reinforcement learning	Reinforcement learning, decision-making through trial and error	DG	Plots typologies, street network generation
Density-based spatial clustering of applications with noise (DBSCAN)	Information clustering, spatial reasoning	PF	Building features identification
Search and optimization	**Search and optimization, problem-solving through exploration**		
Genetic algorithm (GA) based AI	Evolutionary algorithms, imitation of natural selection	DG	Block massing optimization, buildings shape optimization
Non-dominated sorting genetic algorithm II	Evolutionary algorithms, multi-objective optimization	DG	Optimal block massing optimization
Quantum annealing-based AI	Quantum optimization, problem-solving through quantum effects	DG	Plots typologies optimization

(continued)

Table 3 (continued)

AI Applications in urban design			
AI algorithms	Heuristic methods	Design process	Scope in urban design
Machine learning (ML)	Data-driven learning and pattern recognition		
Knowledge-based system	**Rule-based reasoning, knowledge integration**		
Valence aware dictionary and sentiment reasoner (VADSR)	Sentiment analysis, emotional reasoning	PF	User pedestrian experience evaluation
Rule-based + case based reasoning	Hybrid reasoning, integrating rules with past experiences	PF	Diversity prediction, Recommendations
Case-based reasoning	Case-based reasoning, learning from past experiences	DG	Case study research, Recommendations
Embodied intelligence	**Embodied learning, integrating sensory information**		
Smart agent based model	Agent-based simulation, modeling human behavior	DE	Pedestrian flow simulation
Cellular automata (CA)-based AI	Pattern generation and simulation, spatial reasoning	DG, DE	Block massing prediction, zone building massing generation, plot building massing generation

Source By Authors, Data derived from: Authors, [4]
PF: Problem formulation
DG: Design generation
DE: Design evaluation

3.1 *Urban Design and Human–Computer Interaction (HCI): Multi-Disciplinary Integration*

AI-driven heuristic inquiry significantly improves urban design approaches, including both evidence-based and automated methods. More crucially, AI technologies empower urban designers and researchers with deeper insights into human spatial behaviors and preferences, promoting a methodological shift that integrates evidence-based form-making and HCI.

Initially, HCI applications in urban design primarily utilized media façades and urban screens to create interactive communication channels and enhance cultural values within streetscapes [30]. Recently, however, HCI research increasingly focuses on understanding and shaping human experiences within built environments

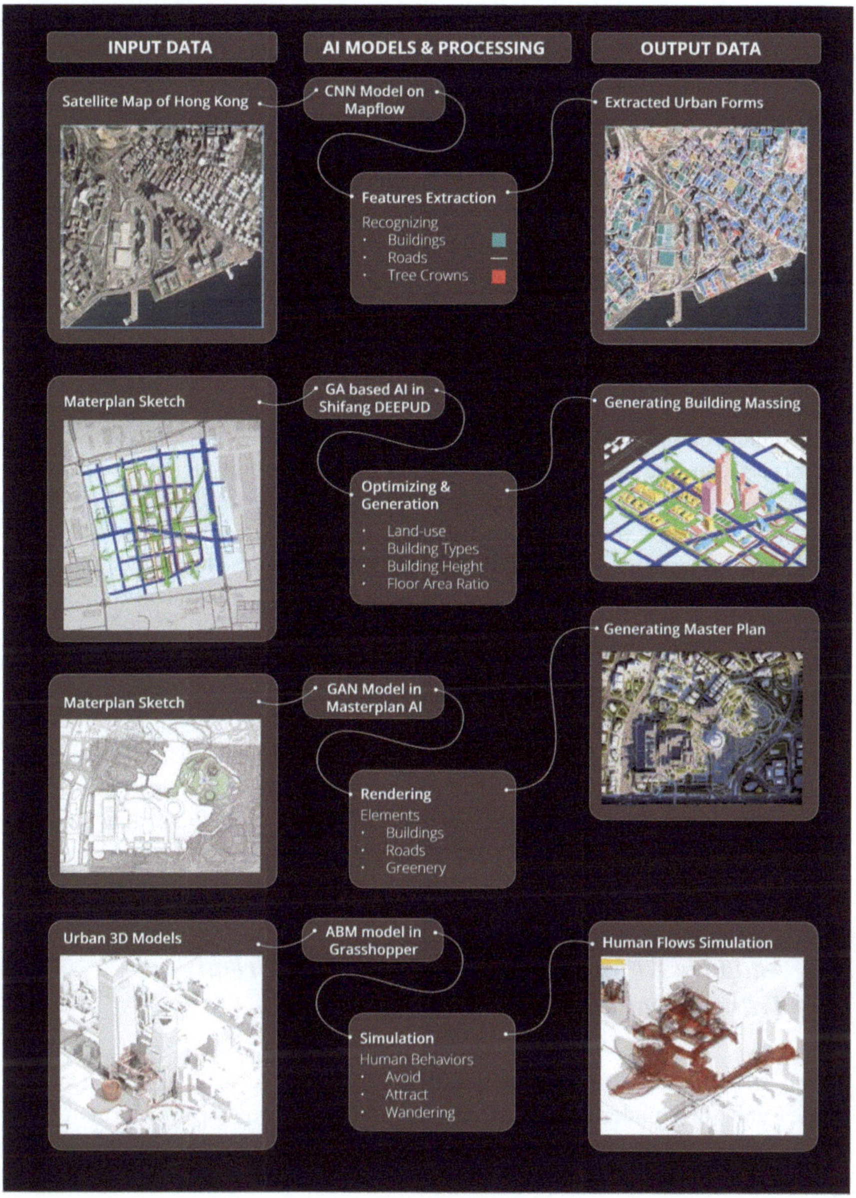

Fig. 2 Example of heuristics AI applied in urban design processes. (*Source* By Authors)

[31]. By leveraging AI's pattern recognition and classification abilities, researchers can better evaluate intricate behavioral patterns and human–environment interactions, significantly expanding HCI applications. Furthermore, interactions within virtual reality (VR) urban environments have emerged as a methodological framework to systematically assess human values, needs, and priorities, addressing the complexities of interactive urban experiences [32, 33].

For instance, Choi and Zhang [34] developed an evaluation framework integrating urban design theory and neuroscience to assess emotional responses elicited by vertical urban complexes under diverse spatial conditions. This methodology employs EEG devices to measure emotional states while experiencing VR simulations, establishing correlations between emotional responses and spatial cognition attributes identified by AI, such as taxonomy, color, depth, boundary, and shape. This structured HCI-based methodology quantifies how vertical urban environments influence human cognition and emotional well-being, demonstrating quantitatively how natural elements like sunlight and greenery, spatial permeability, legibility, and accessibility improve emotional experiences in high-rise urban settings [32]. Utilizing this structured evaluation framework, urban designers could create perceptually responsive urban environments enhanced by IoT and smart sensors. These developed urban environments might not only able to capture real-time data but also to understand and respond to human emotions and behaviors.

To effectively address human-cantered social issues in urbanism, urban design naturally requires multidisciplinary methodological integration. These transformative methodological shifts, integrating neuroscience, psychology, and anthropology, are poised to form new ideologies akin to the evolution of modernism. This multidisciplinary approach enables the design of responsive urban spaces that actively engage with human perceptions, emotions, and potential social conflicts, fundamentally redefining contemporary urban experiences.

3.2 *Adaptive AI Models for Data-Informed Urban Transformation*

Besides, AI-aided urbanism introduces structural transitions in urban design methodologies and stakeholder dynamics. AI technology effectively transforms qualitative urban design challenges into quantitative analysis and models, significantly enhancing reliability, effectiveness, and precision in current design methods [4, 35, 36]. While these advancements benefit designers, the significance of AI technology lies in the emergence of data-informed urban design models—where both quantitative data (e.g., big data, user-generated data) and qualitative data (e.g., semantic inputs) are leveraged throughout the design process. This includes informing conceptual frameworks, guiding ideation, and supporting decision-making [37].

This shift is particularly evident in the increasing integration of AI into design tools, enabling the extraction of insights from mixed data sources and facilitating

precise alignment with local conditions. In practice, many technology firms and design consultancies are developing proprietary AI models specifically tailored to local contexts, using data from diverse networks and sources [7, 38]. A notable example is Zencity, an Israeli govtech firm active globally. Zencity provides an AI-driven community engagement platform widely adopted by local governments. It aggregates real-time resident feedback from social media, local news, city hotlines, and online surveys, analyzing data using proprietary natural language processing to categorize topics and gauge public sentiment [39]. Its ability to synthesize mixed data sources and deliver hyper-localized insights has positioned Zencity as a critical intermediary between designers and communities, enabling more responsive and targeted urban development.

As a result, urban design firms and data-driven consultancies that leverage proprietary AI methodologies are reshaping market dynamics. By combining AI with extensive local datasets and insights, these data professionals achieve competitive advantages unattainable by generic digital tools or conventional consultants [40]. Consequently, local governments increasingly seek collaboration with AI-driven specialists capable of delivering highly contextualized, predictive solutions. **In this evolving AI era, the constellation of essential stakeholders now includes data collection companies, platform providers, data engineers, and analysts—playing roles as vital as those of architects, urban designers, and engineers**.

This restructuring of stakeholder roles—both individual agents and institutional agencies—significantly redefines responsibility for urban transformation across a five-stage, data-informed design workflow (Table 4). In the early stages of design, data analysts, data service providers, and civic-tech facilitators play a pivotal role by assembling baseline evidence and framing human-centered design goals. As the project advances into concept testing and optimization, computational designers, data scientists, sensor technicians, and operations research specialists take the lead, collaborating closely with the core design team. In the final stage, data analysts and data engineers re-engage to conduct evaluations, weigh trade-offs, and support decision-makers in approving optimized design schemes. This progression—from broad diagnostic input to specialized technical and evaluative expertise—characterizes the professional shift within the data-informed design model.

Ultimately, AI-aided urbanism enables the emergence of new design models that prioritize adaptability, cross-sector collaboration, and real-time responsiveness, contributing to more dynamic, data-informed processes of urban transformation.

3.3 Morphological Vision

Unlike the large-scale expansion into megacities seen in recent decades (e.g., the great bay area in China, Tokyo, Delhi, and Mumbai), the most significant developments in urban form through digital sensors and heuristic AI tools are not occurring at the scale of city-wide growth. Rather, they focus on optimizing efficiency within the existing urban fabric, responding to local features and constraints [41]. AI-aided

Table 4 Mixed data sources and data professionals of data-informed design across five urban design process

Urban design process	Purposes of data-informed design	Examples of mixed data sources for AI analysis	Key data professionals
1 Initial conceptualization 2 Ideation and concept development	Diagnose existing gaps; set evidence-based project goals	• GIS layers (street network, land-use, transit, crash maps, demographic and health stats) • Open-data portals • Community surveys and perception audits on social media • Participatory GIS/ mobile phone dataset • Sensors data (Wi-Fi/ Bluetooth, CCTV)	• Data analysts • Civic-tech companies • Data-service companies
3 Prototyping (high-fidelity sim using VR/AR)	Measure performance of draft design; capture behavioral, safety and health effects before full build-out	• Temporary IoT sensors • Video analytics and computer-vision data • Social media for live feedback	• Data engineers / IT technicians deploying sensors • Data labs processing mobile and video data • Civic-tech teams running feedback apps
4 Iteration and refinement	Optimize layout, program, or operations until metrics improve	• Time-series sensor and video datasets from pilots • Updated community feedback • Maintenance/ cost-tracking data	• Data analysts interpreting pilot metrics • Optimization/ operations-research specialists • Budget/feasibility analysts
5 Decision-making	Supporting decision-making by predictive modeling or data-driven validation	• Multi-criteria decision matrices and scoring	• Data analysts/data engineers

Source By Authors

urbanism drives morphological changes at two distinct levels: (1) the creation of new urban form typologies, and (2) the enhancement of efficiency in urban transformation processes.

As previously discussed, AI's primary influence lies in reshaping urban design methodologies and workflows rather than directly generating new typology of built forms. However, its impact on the built environment should not be underestimated. Unlike passive digital sensors embedded in infrastructure, AI interacting closely with users would gradually stimulate typological shifts in spatial design, enhancing human interactions and spatial experiences [6, 42]. Substantially, AI-driven spatial HCI also significantly redefines architectural temporality and functionality, addressing the tension highlighted by Johansen et al. [43] between durable architectural outcomes and transient interactive experiences. Accordingly, AI-driven urbanism promotes urban forms characterized by adaptability, functional fluidity, and significant integration with digital and interactive elements.

These interactive and real-time responsive characteristics inherently enhance the efficiency with which urban forms address diverse user needs. For example, Mubeen et al. [44] demonstrate an adaptive marketplace in Bhopal's 10 Number Market, where ceiling-mounted cameras and occupancy sensors inform a central AI system that reconfigures vendor layouts hourly. In the morning, modular kiosks cluster near transit stops to facilitate quick grocery shopping. By afternoon, approximately 40% of these units reposition to perimeter zones, forming pedestrian promenades with adaptive seating that transiting from benches to dining areas based on detected user postures. Such innovative transformations are increasingly evident in public spaces like bus stops, plazas, and streetscapes, dynamically adjusting seating arrangements or protective canopies in response to real-time user patterns [45, 46]. As a result, researchers and designers refer to these adaptive spaces as "Cognitive Urban Interfaces," highlighting their ability for autonomous spatial reconfiguration and decision-making [2, 44, 47].

Building upon these insights, Choi [48] introduces the concept of "urban hypermorphology," envisioning urban forms characterized by bottom-up processes, self-organization, and real-time adaptability to environmental dynamics and user needs. Drawing inspiration from biological principles, this approach proposes innovative strategies for designing multi-layered, user-centric urban environments capable of responding dynamically to complex human behaviors and interactions. To actualize this vision, the aesthetic and functional integration of digital components, such as sensors, screens, and interactive interfaces, is crucial. Sustainable energy solutions, including solar panels, should also be reinterpreted as essential spatial elements, facilitating greater integration and visual coherence.

This new concept offers a robust framework for understanding urban dynamics and fostering interactive communities by examining how spatial systems respond to local cultural contexts, social dynamics, and human activities while reflecting the unique interplay of social norms, values, and built environments. Distinct from traditional urban development processes through research, design, and implementation (outlined in Fig. 1), AI-aided urbanism enables morphological transformations independent of constant designer intervention. Utilizing AI-driven reasoning and real-time data analysis, this comprehensive approach significantly reshapes and continually adapts urban landscapes, optimizing efficiency and responsiveness to evolving urban demands.

4 Discussion: Toward Inclusive Urbanism

In the previous section, the chapter demonstrated current trends in design disciplines and morphological transformations driven by AI applications. The emerging design models, integrating human–computer interaction (HCI) disciplines and data-informed design processes, strive toward inclusive urbanism that welcomes and accommodates diverse human needs—ensuring that individuals of all ages, abilities, genders, cultures, and socioeconomic statuses feel comfortable and valued in public spaces. AI-driven HCI methodologies significantly enhance inclusivity by facilitating broader community engagement. These approaches democratize design participation, empowering residents—especially underserved groups—to express spatial preferences and emotional needs through accessible, real-time feedback mechanisms.

Moreover, cognitive urban spaces enabled by heuristic AI inherently support inclusivity by dynamically addressing diverse accessibility needs. Examples such as modular kiosks and adaptive seating optimize spatial efficiency and substantially enhance usability for elderly residents, individuals with disabilities, and families when developed through detailed demographic analysis (see Sect. 3.3). These inclusivity-driven morphological innovations reinforce equitable access, demonstrating AI's potential to create urban environments universally designed for diverse populations.

Building upon these advancements, inclusive urban design is becoming a practical and tangible discipline in urban planning. Machine learning models can integrate data on land use, transit systems, demographics, and services to identify neighborhoods underserved by essential facilities like parks and clinics or highlight infrastructure gaps, such as missing sidewalks or safe crossings in lower-income areas. Social impact prediction is another emerging application, where socio-economic data, housing trends, and travel patterns are input into AI systems to forecast the social implications of urban development plans [49]. By anticipating these impacts, planners can proactively adjust designs to mitigate potential harms—for example, by including affordable housing or enhanced accessibility features. Such methodologies illuminate the specific needs of groups that are often statistically "invisible" in standard metrics, including people with disabilities, whose travel patterns might be numerically limited yet significantly distinct, or minority cultural groups whose spatial usage patterns are typically underrepresented in mainstream surveys.

Recognizing these capabilities, AI technologies support inclusive urban design by revealing hidden needs, amplifying community input, and predicting the consequences of design choices. By handling extensive data analyses, AI allows urban planners to concentrate on nuanced values and ethical judgments beyond the scope of computational abilities—such as ethically balancing competing community interests.

However, the practical application of these technologies is not without its challenges and risks. Collecting sufficient high-quality and wide demographic data to improve AI performance is both costly and time-consuming. Although data firms and infrastructure providers contribute to localized databases, scattered and fragmented datasets hinder unified integration, complicating their adaptation within design workflows.

Governance challenges surrounding AI use also present significant concerns for inclusive urbanism. Autonomous AI systems shift some decision-making powers from humans to machines, raising transparency and accountability issues due to AI's inherent opacity—the "black-box effect" [28]. This opacity risks compromising fairness and alignment with human values, exacerbated by current performance limitations. Effective governance is essential to manage biases, misalignment, and accountability uncertainties, yet legal and regulatory frameworks remain inadequate [2]. Nevertheless, these technical and governance challenges may be progressively addressed through technological and social advancements. Ultimately, the most profound challenges for inclusive urbanism in the AI era fundamentally revolve around ethical considerations in data collection and fragmented data ownership associated with data-driven urban development.

4.1 *Ethical Data Usage in HCI-Based AI Urbanism*

In smart cities, data collection traditionally targets environmental monitoring and urban infrastructure management. However, the shift toward AI-aided urbanism elevates human-centric data collection, especially crucial for designing interactive urban spaces. Current practices involving social media, mobility applications, and cellular services already collect substantial user data. Real-time data collection methods, such as facial recognition technologies in public spaces, pose significant ethical dilemmas related to privacy, consent, and security. Transparency in data lifecycle management, particularly within public administration and private sectors providing public services, is critical to maintaining public trust and community engagement [50]. Implementing transparent processes that allow citizen oversight or recourse is essential for responsibly managing future ubiquitous data collection.

Beyond data collection, cybersecurity poses another critical concern. The vast volumes of sensitive personal data from sensors and surveillance systems represent attractive targets for misuse or cyberattacks. Recent research highlights escalating cybersecurity breaches—for instance, data breaches in the U.S. increased by 170% from 2010 to 2021, with average breach costs rising to $9.44 million per incident [35]. Cities must thus implement robust cybersecurity measures, including data encryption, controlled access, and regular audits, alongside compliance with privacy regulations. Balancing innovation with stringent security requirements remains a crucial challenge, demanding transparent and respectful utilization of real-time human data in future HCI-driven urbanism.

4.2 Data Ownership and AI-Aided Urban Design

Urban data ownership frequently remains fragmented across multiple governmental levels and between public and private actors, resulting in conflicts and inefficiencies. Disputes over multi-level governance arise when municipal, regional, and national authorities assert competing claims over urban-generated data. Overlapping or inconsistent regulations, ranging from national privacy laws to local data policies, increase compliance costs and can stall urban initiatives. Excessive or uncoordinated regulatory frameworks can severely limit cities' abilities to effectively gather and share essential data [51]. Clearly defining roles and enhancing coordination among stakeholders is crucial to resolving these complexities. Innovative data governance frameworks, such as data trusts or cooperative data platforms that involve diverse stakeholders, have been proposed to address these challenges [2].

Conflicts also emerge between public authorities and private firms regarding data ownership. Many urban data streams—from sensors and cameras to telecom networks—are operated by private companies through contracts or public–private partnerships. This can result in data silos or proprietary gatekeeping practices, with private companies treating urban data as commercial assets. Such opaque data governance negatively impacts public interest goals, including inclusivity and innovation [52]. However, a fragmented, siloed approach, where each company or city department manages data in isolation, might constrain designer to work around data fiefdoms, undermining the interoperability needed for creative urban design. To counter this, data ownership standards and collaborative governance need to address this dilemma problem.

5 Conclusion

This chapter adopted a systematic review methodology, supplemented by comparative analysis and case study exploration, to critically examine the evolving roles and values of AI in shaping contemporary urbanism. Through a socio-technical lens, the review traced historical technological transformations—such as industrialization, CAD, and digital urbanism—and positioned AI-aided urbanism as a paradigmatic shift. Unlike previous innovations, AI introduces heuristic, data-driven, and real-time responsive capabilities that embed themselves across all layers of urban development. As illustrated in Fig. 3, the evolution of technology development in relation to urban development—reflected in its changing forms and characteristics—has been closely tied to advancements in design and construction technologies. From industrialization and internationalism to CAD-enabled mass production, symbolized by towering buildings and mega-structure, and finally to smart cities and ultimately digital twins, each wave has left a distinct imprint on the urban fabric. In this context, the chapter highlights how AI differs not merely in efficiency or automation but in its potential to redefine the very paradigm of urbanism—through its influence on design disciplines, procedural models, stakeholder roles, and the resulting spatial and morphological expressions of our cities.

In conclusion, these insights affirm that AI-aided urbanism is not merely a technical enhancement but a foundational transformation of urban design, governance, and spatial expression. It offers a critical opportunity to reframe urban development

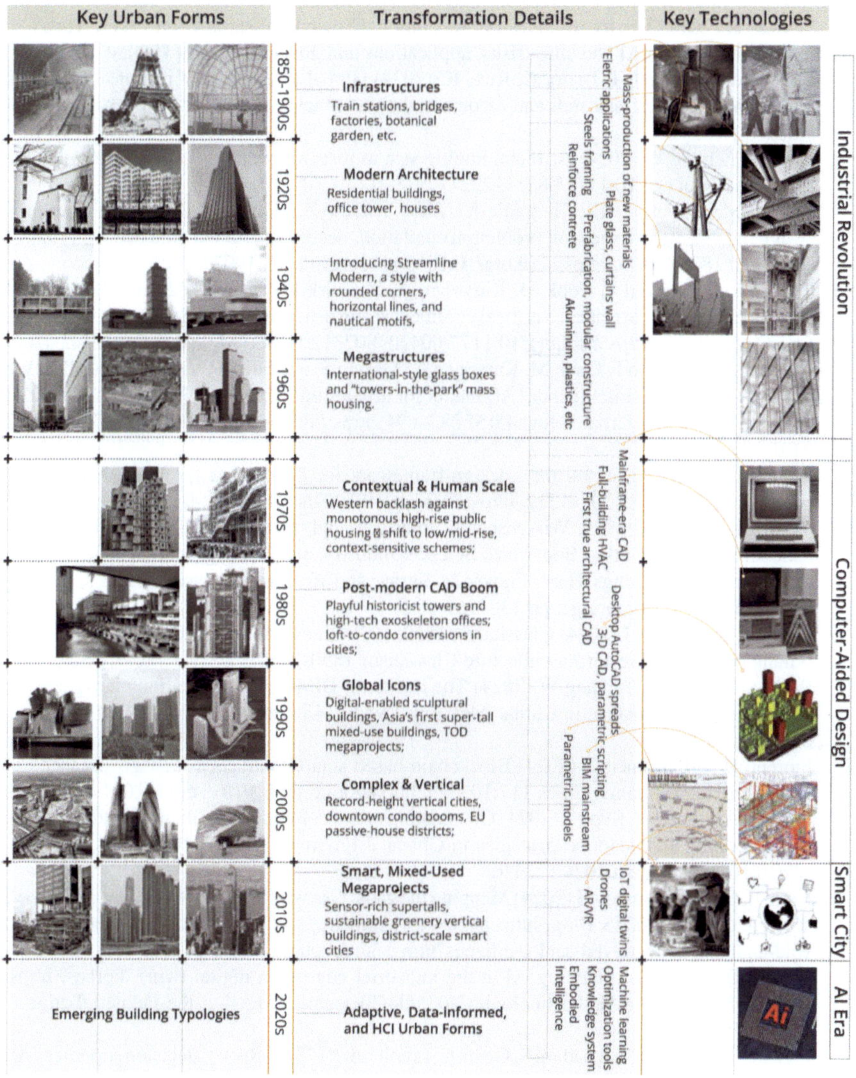

Fig. 3 Evolution timeline of technology development related to urban building typologies (*Source* By Authors. Note. images within the figure are under Creative Commons license. Full bibliographic entries are listed in Reference section.)

as an inclusive, participatory, and adaptive practice responsive to the complexities of twenty-first century cities.

References

1. Schwab K (2017) The fourth industrial revolution. Crown Publishing Group, New York
2. UN-Habitat (2022) AI and cities risks, applications and governance. UN-Habitat, Nairobi
3. Ben Dhaou S, Isagah T, Distor C, Ruas IC (2024) Global assessment of responsible artificial intelligence in cities: research and recommendations to leverage AI for people-centred smart cities
4. Choi HS, Zhang W (2024) Artificial intelligence as research methods in urban design. J Plan Liter https://doi.org/10.1177/08854122241291820
5. Jiang F, Ma J, Webster CJ, Chiaradia AJ, Zhou Y, Zhao Z, Zhang X (2024) Generative urban design: a systematic review on problem formulation, design generation, and decision-making. Prog Plan 180:100795. https://doi.org/10.1016/j.progress.2023.100795
6. Cugurullo F, Caprotti F, Cook M, Karvonen A, McGuirk P, Marvin S (2023) The rise of AI urbanism in post-smart cities: a critical commentary on urban artificial intelligence. Urban Stud 61(6):1168–1182. https://doi.org/10.1177/00420980231203386(Originalworkpublished2024)
7. Caprotti F, Cugurullo F, Cook M, Karvonen A, Marvin S, McGuirk P, Valdez AM (2024) Why does urban Artificial Intelligence (AI) matter for urban studies? Developing research directions in urban AI research. Urban Geogr 45(5):883–894. https://doi.org/10.1080/02723638.2024.2329401
8. Bratton B (2021) AI urbanism: a design framework for governance, program, and platform cognition. AI Soc 36:1307–1312. https://doi.org/10.1007/s00146-020-01121-9
9. Gottfried S (1852) 1989 Wissenschaft, Industrie und Kunst: Vorschläge zur Anregung nationalen Kunstgefühles. Eng. trans: Science, Industry, and Art. The four elements of architecture and other writings, Harry Francis Mallgrave and Wolfgang Herrmann, trans. Cambridge University Press, Cambridge, pp 130–167
10. Rice DT, Boardman J (2024) Classicism, 1830–1930. Encyclopædia Britannica. https://www.britannica.com/art/Western-architecture/Classicism-1830-1930
11. Wang S, Chong H-Y, Zhang W (2024) The impact of BIM-based integration management on megaproject performance in China. Alex Eng J 94:34–43. https://doi.org/10.1016/j.aej.2024.03.036
12. Fiorentino S, Bartolucci S (2021) Blockchain-based smart contracts as new governance tools for the sharing economy. Cities 117:103325. https://doi.org/10.1016/j.cities.2021.103325
13. Ibrahim A, Kuuire V, Kepe T (2024) On mapping urban community resilience: land use vulnerability, coping and adaptive strategies in Ghana. J Environ Manage 370:122426. https://doi.org/10.1016/j.jenvman.2024.122426
14. Bulut E, Goodale-Sussen G (2024) Mapping the gaps: Use void analysis to uncover underserved senior citizens. ArcGIS Blog. https://www.esri.com/arcgis-blog/products/bus-analyst/health/void-analysis-underserved-senior-citizens?utm_source=chatgpt.com
15. Chused B (2025) Accelerating AI at the industrial edge with digital twins. Forbes. https://www.forbes.com/sites/delltechnologies/2025/04/30/accelerating-ai-at-the-industrial-edge-with-digital-twins/
16. Pignatelli M, Torabi Moghadam S, Genta C, Lombardi P (2023) Spatial decision support system for low-carbon sustainable cities development: an interactive storytelling dashboard for the city of Turin. Sustain Cities Soc 89:104310. https://doi.org/10.1016/j.scs.2022.104310
17. Remondino F (2011) Heritage recording and 3D modeling with photogrammetry and 3D scanning. Remote Sensing 3(6):1104–1138. https://doi.org/10.3390/rs3061104
18. Mehan A, Mostafavi S (2024) Emerging technologies in urban design pedagogy: augmented reality applications. ARIN 3:29. https://doi.org/10.1007/s44223-024-00067-y

19. Abdullayeva Y (2024) Applications of parametric design in urban planning. Parametric Architecture. https://parametric-architecture.com/applications-of-parametric-design-in-urban-planning/?srsltid=AfmBOoqmABgB6o4hMLwD2X2BWU_s-L9HYzelxz7hKM57z8XmwcfY1219&utm_source=chatgpt.com
20. Yusuf O, Rasheed A, Lindseth F (2025) Exploring urban mobility trends using cellular network data. In: Kioumarsi M, Shafei B (eds) The 1st International conference on net-zero built environment. NTZR 2024. Lecture notes in civil engineering, vol 237. Springer, Cham. https://doi.org/10.1007/978-3-031-69626-8_138
21. Pascoli PR, Bezerra MDM (2024) The digital twin in urban planning and management: a key asset in addressing the challenges presented by on-going climate changes. In: Gauthier S, Nicol F, Brotas L, Altamirano H (eds.) 13th masters conference: people and buildings. London, United Kingdom, pp 1–6. https://doi.org/10.5258/SOTON/P1193
22. Zaman M, Puryear N, Abdelwahed S, Zohrabi N (2024) A review of IoT-based smart city development and management. Smart Cities 7(3):1462–1501. https://doi.org/10.3390/smartcities7030061
23. Partida D (2024) Leveraging AI Chatbots to enhance citizen engagement in City Services. Planetizen Blogs. https://www.planetizen.com/blogs/131448-leveraging-ai-chatbots-enhance-citizen-engagement-city-services?utm_source=chatgpt.com
24. Höffken S, Streich B (2013) Mobile participation: citizen engagement in urban planning via smartphones. In: Silva C (ed.) Citizen E-participation in urban governance: crowdsourcing and collaborative creativity. IGI Global Scientific Publishing, pp 199–225. https://doi.org/10.4018/978-1-4666-4169-3.ch011
25. Han JH, Hawken S, Williams A (2015) Smart CCTV and the management of urban space. In: Harrison D (ed) Handbook of research on digital media and creative technologies. IGI Global, Hershey, PA, pp 430–447. https://doi.org/10.4018/978-1-4666-8205-4.ch019
26. Zhao H, Niu C, Dou X, Liang J (2024) Urban multipoint fire disaster emergency simulation based on web information. Int J Disast Risk Reduct 101:104223. https://doi.org/10.1016/j.ijdrr.2023.104223
27. Najafi H, Tudor K, Ings W (2024) A review of the evolution of heuristic inquiry Heuristic Enquir 9–22 https://doi.org/10.4324/9781003507758-3
28. Ghobakhlou A, Najafi H (2024) A comparative study of heuristic inquiry in AI and artistic research. Heuristic Enquir 139–161. https://doi.org/10.4324/9781003507758-12
29. Lartey D, Law KM (2025) Artificial intelligence adoption in urban planning governance: a systematic review of advancements in decision-making, and policy making. Landsc Urban Plan 258:105337. https://doi.org/10.1016/j.landurbplan.2025.105337
30. Fischer PT, Hornecker E (2012) Urban HCI: spatial aspects in the design of shared encounters for media façades. In: Proceedings of the SIGCHI conference on human factors in computing systems. pp 307–316. https://doi.org/10.1145/2207676.2207719
31. Alavi HS, Churchill EF, Wiberg M, Lalanne D, Dalsgaard P, Fatah gen Schieck A, Rogers Y (2019) Introduction to human-building interaction (HBI). ACM Trans Comput-Hum Interact 26(2):1–10. https://doi.org/10.1145/3309714
32. Choi HS, Bruyns G, Zhang W, Cheng T, Sharma S (2023) Spatial cognition and three-dimensional vertical urban design guidelines—cognitive measurement and modelling for human centre design. Urban Sci 7(4):125. https://doi.org/10.3390/urbansci7040125
33. Marcus W, Paay J, Langenheim N, Tianyi Y (2025) HCI methods supporting urban design evaluation using virtual environments. Interact Comput. https://doi.org/10.1093/iwc/iwaf014
34. Choi HS, Zhang W (2025) The way to measure and establish an emotional-based assessment of Vertical Urban Complex. Cities 163:106015. https://doi.org/10.1016/j.cities.2025.106015
35. Aslam M, Abbasi MAK, Khalid T, Shan RU, Ullah S, Ahmad T, Saeed S (2022) Getting smarter about smart cities: Improving data security and privacy through compliance. Sensors 22(23):9338. https://doi.org/10.3390/s22239338
36. He W, Chen M (2024) Advancing urban life: a systematic review of emerging technologies and artificial intelligence in urban design and planning. Buildings 14(3):835. https://doi.org/10.3390/buildings14030835

37. Quiñones-Gómez JC, Mor E, Chacón J (2024) Data-driven design in the design process: a systematic literature review on challenges and opportunities. Int J Hum-Comput Interact 41(4):2227–2252. https://doi.org/10.1080/10447318.2024.2318060
38. Huang X, Zeng L, Liang H et al (2024) Comprehensive walkability assessment of urban pedestrian environments using big data and deep learning techniques. Sci Rep 14:26993. https://doi.org/10.1038/s41598-024-78041-x
39. Simon M (2025) Zencity named to Fast Company's 2025 list of the world's most innovative companies. Global Press Release & Newswire Distribution Services. https://www.businesswire.com/news/home/20250318461201/en/Zencity-Named-to-Fast-Companys-2025-List-of-the-Worlds-Most-Innovative-Companies
40. Funk M, Lovei P, Noortman R (2024) Designing with data, data-enabled and data-driven design. In: Vanderdonckt J, Palanque P, Winckler M (eds) Handbook of human computer interaction. Springer, Cham. https://doi.org/10.1007/978-3-319-27648-9_40-1
41. Batty M (2008) The size, scale and shape of cities. Sci 319:769–771
42. Strengers Y, Kennedy J (2020) The smart wife: why siri, alexa, and other smart home devices need a feminist reboot. MIT Press, Cambridge, MA
43. Johansen SS, Kjeldskov J, Skov MB (2019) Temporal constraints in human–building interaction. ACM Trans Comput-Hum Interact 26(2):1–29. https://doi.org/10.1145/3301424
44. Mubeen A, Paliwal A, Soni S (2025) Introducing IOT in the realm of public space—a case study of Bhopal. Int J Sci Dev Res 10(1):a746–a755. https://ijsdr.org/papers/IJSDR2501078.pdf
45. MIT (2018) Interactive bus stop. http://mobile.mit.edu/projects/redefning-urban-bus-stops/. Accessed on 02 Mar 2018
46. George M (2025) 10 Inspiring interactive public space designs worldwide. UDL Education Pvt. Ltd. https://urbandesignlab.in/10-inspiring-interactive-public-space-designs-worldwide/
47. West P (2025) The Thinking City: building the living urban ecosystem of tomorrow. Electronic Specifier. https://www.electronicspecifier.com
48. Choi HS (in press) Human-centred design: spatial cognition and hyper-morphology. Springer Nature
49. Mahendra S (2023) Artificial Intelligence and urban design. Artificial Intelligence. https://www.aiplusinfo.com/blog/artificial-intelligence-and-urban-design/#:~:text=AI%E2%80%99s%20ability%20to%20analyze%20large,indicator%20of%20social%20acceptance%20or
50. Sanfilippo MR, Frischmann B (2023) Slow-governance in smart cities: an empirical study of smart intersection implementation in four US college towns. Internet Policy Rev 12(1). https://doi.org/10.14763/2023.1.1703
51. Organisation for Economic Co-operation and Development (OECD) (2023) Smart city data governance: challenges and the way forward (OECD Urban Studies series). OECD Publishing
52. Franke J, Gailhofer P (2021) Data governance and regulation for sustainable smart cities. Front Sustain Cities 3. https://doi.org/10.3389/frsc.2021.763788

Figure 3 Bibliographic Entries

53. Atomic Taco (2013) AutoCAD, AutoDesk 1986, Licensed under CC BY-SA 4.0
54. Bradley B (2010) Fixing-tilt-slabs. Licensed under CC BY-SA 3.0
55. Bundesarchiv (1959) Bild 183–62915–0005. VEB Stahl- und Walzwerk Brandenburg, Licensed under CC BY- NC-SA 3.0
56. CC BY-SA 4.0
57. Choo YS (2015) Park Royal Tower. Licensed under CC BY-SA 2.0
58. DigbyDalton (2018) Hudson yards from the empire state building. Licensed under CC BY-SA 3.0
59. EG Focus (2011) Westfield stratford city and olympic park CGI. Licensed under CC BY 2.0

60. Exploringlife (2016) Kowloon Station in West Kowloon, Hong Kong. Licensed under CC BY-SA 2.0
61. GauloisUSA (1986) The world trade center seen from the Hudson, New Jersey shore (northwest). Licensed under CC BY-SA 4.0
62. Iwan Baan (2014) Zaha Hadid—香港理工大學賽馬會創新樓. Licensed under CC BY-SA 2.0
63. Jernej F (2023) AI Artificial Intelligence concept. Licensed under CC BY-SA 4.0
64. Williams J (2009) Bauhaus home 1929. Licensed under CC BY-NC-SA 2.0.
65. Ohyama K (2017).The seagram building, viewed from street level, from Park Avenue. Licensed under CC BY-SA 4.0
66. Museum, The Board of Trustees of the Science (2023) Nasmyth's steam hammer. World History Encyclopedia. https://www.worldhistory.org/image/17048/nasmyths-steam-hammer/, Licensed under CC BY- NC-SA 4.0.
67. Howard N (2015) The Gherkin in London at Dawn. Licensed under CC BY-SA 3.0
68. Pacific Place (2011) 3d map of pacific place Hong Kong. Licensed under CC BY-SA 2.0
69. Fussan S (2016) Shanghai Tower. Licensed under CC BY-SA 3.0
70. Parnell T (2022) Barbican. Licensed under CC BY-SA 2.0
71. Wing (2008) HK HSBC main building. Licensed under CC BY 3.0
72. Проектирование трубопроводов (2020) в T-FLEX CAD. https://www.tflex.ru/. Licensed under CC BY-SA 2.0

The Role of Digital HRM in Tech-Enabled Urbanism: Navigating Workforce Challenges in Smart Cities

Rajwinder Kaur and Anmol Sharma

Abstract With smart cities transforming the landscape of urban governance, infrastructure, and citizen engagement, Human Resource Management (HRM) must move away from the organizational efficiency paradigm and embrace human-centered, inclusive, and sustainable workforce ecosystems. The chapter considers digital HRM which includes tools such as AI-enabled recruitment, predictive analytics, and decentralized systems—as a means to be a strategic enabler in tech-enabled urban settings. It is seen that Digital HRM not just as isolated support function of other HR practices but at the core of workforce transformation in smart cities, specifically in terms of bridging digital skills gaps, including gig workforce members, and ethical technology governance. The study considers Digital HRM through the lenses of RBV, TAM, and Institutional Theory, synthesizing interdisciplinary literature on workforce challenges associated with technology-enabled contexts, particularly displaced employment, mental well-being, and equity. Moreover, it includes real-world examples from global smart cities (i.e., Singapore, Amsterdam, Dubai) as well as technology firms (i.e., Google, Microsoft, IBM) to illustrate how strategic Digital HRM practices can provide insight into the unique emphasis on adaptive, ethical, and innovative urban labor systems. Finally, the study proposes a Digital HRM outline highlighting people within Digital HRM paradigm that aligns with the goals of innovation, growth and sustainable urban environments while progressing the overarching purpose of comprehensive smart city ecosystems.

Keywords AI in HRM · Digital HRM · Urbanization · Sustainable development goals · Sustainable workforce strategies · Tech-enabled urbanism · Smart cities · Urban planning

R. Kaur (✉)
Apex Institute of Management, Chandigarh University, Mohali, Punjab, India
e-mail: bimbrarajwinder@gmail.com

A. Sharma
Department of Management, Chandigarh School of Business, Chandigarh Group of Colleges, Jhanjeri, Punjab, India

V. Chand Sharma et al. (eds.), *Tech-Enabled Urbanism and Entrepreneurship for Inclusive Cities*, Lecture Notes in Civil Engineering 779,
https://doi.org/10.1007/978-981-95-3873-7_3

1 Introduction

1.1 Tech-Enabled Urbanism and Smart Cities

The global movement toward smart cities constitutes a significant change in how urban places are designed, governed, and engaged with. Smart cities scale from the small neighborhood to the larger city region, integrating physical infrastructure, digital technologies, and human systems to deliver data-responsive and porous environments for the purposes of enhancing quality of life, sustainability, and efficiency. Smart cities are not simply digital advances or upgrades to local civic infrastructure, but rather indications of a systemic transformation to urban ecosystems that is facilitated by a flow of information, citizen engagement, and AI [1, 14]. This transformation is well underway by governments around the world, especially those in emerging economies such as India. The Smart Cities Mission [34] is one of the most ambitious attempts to reorder urban governance and urban services using technology, while also making space for inclusive, sustainable, and citizen-driven urban environments.

1.2 Argument and Central Research Question

While Digital HRM may seem to play a side role as part of a smart city and its functionality, this chapter argues that it is a strategic enabler of an equitable workforce, workforce innovation, and workforce resilience. As cities become increasingly seen as complex sociotechnical systems, HRM must evolve from traditional transactional roles into transformational, data-driven, and ethical modes of workforce governance. The shift to include Digital HRM capabilities in the smart city ecosystem does not only relate to the implementation of technology, but rather the implications of inclusive urban development, digital justice, and socio-economic sustainability.

This chapter examines the guiding research question: How are Human Resource Management and Digital HRM able to strategically align to enable human development, entrepreneurial inclusion, and workforce sustainability in smart city ecosystems?

This discussion is framed as an assumption that Digital HRM can serve as a bridge between digital infrastructure and systems of inclusive workforces for social goods, if Digital HRM is designed ethically and with human-centered design principles.

1.3 Research Objectives

1. To analyze how digital transformation is reshaping HRM paradigms in smart cities, moving beyond automation toward strategic and inclusive HR models.

2. To examine the workforce challenges that arise within tech-enabled urban environments.
3. To propose a strategic, human-centric strategies' that aligns digital HRM capabilities with the long-term sustainability and inclusivity goals of smart cities.

1.4 Significance and Novel Contribution

This chapter provides a new scholarly contribution to two vibrant bodies of research, smart urbanism and digital HRM, both of which tend to be examined separately in the academic literature. This chapter contributes to a new but important area of research at the nexus of "smart" urbanism and workforce development. Numerous scholars have examined the technological, infrastructural, and realized material aspects of smart urbanism [5, 21], while HR scholars have examined corporate digital HR transformation [12, 37]. However, very few academic works have closely examined smart urbanism by coalescing both urbanism and Digital HRM to examine the strategic role of Digital HRM in the continued creation of smart, inclusive cities.

Finally, this chapter calls for HRM perspectives to be embedded in urban planning processes, and contends that human capital and digital HR strategies need to be the principal consideration in the smart city visions adopted by urban planners and leaders. In this, it offers a novel theoretical and policy-recognitive intervention to more closely align digital transformation with sustainable, equitable urban futures.

2 Background and Literature Review

Through the combination of new digital infrastructures, data, and human services these smart cities are changing the urban landscapes. Transforming cities into smart, visible enterprises not only requires technological advancement but also a transformation in workforce management. Although many smart city projects and initiatives focus on innovative services, connectivity, and sustainability, successful projects set out to take the human factor into account. The study proposes that digital HRM can be a strategically centric aspect of sustainable, inclusive, and entrepreneurial workforce ecosystems in smart cities.

2.1 From Traditional HR to Digital HRM

The movement from Traditional HRM to Digital HRM is a movement from manual, paper-driven processes to integrated, technology-enabled tools and processes. Digital HRM (often referred to as e-HRM) will be understood in this paper as the use of web-based data and technology to conduct many of the HR functions required, to conduct employee recruitment, performance management, and employee engagement [29]. Digital HRM began fundamentally as a plug-in that would be more efficient in improving administrative processes, and it thus evolved into an organizational resource that can transform organizational cultures and adapt workforces [12]. In contexts of smart cities-where complexity happens- HRM must go beyond operational HRM and be a mechanism for change. Study commented that strategic e-HRM supports agility and innovation, such that real-time access to data, access to cross-functional processes, and analytic capabilities related to people are important in smart, urban labor dynamics [37].

Smart cities represent a paradox: centers of innovation, but also typically afflicted by skills mismatches, workforce inequality and labor segmentation [5]. The fourth industrial revolution is both a challenge and an opportunity for HRM systems and organizations as they begin to understand and appreciate the radical changes in work across humanity as articulated by the Extant studies argue that the highly skilled knowledge workers of the future will need knowledge related to digital literacy, complex problem solving, and socio-emotional intelligence [31, 52]. Study estimates globe-wide up to 14% of the total global workforce, which is about 375 million workers, will need to re-skill or switch roles to keep pace with the rapid pace of change by 2030 [31]. While the disruptions present challenges, there are unique opportunities for Digital HRM, particularly with respect to future skills demand, upskilling and continuing education, and adaptable workforce structures [41]. Via Learning Management Systems (LMS), mobile applications, and AI-enabled career pathing, HRM professionals can close workforce skills gaps, while providing unique and personal learning and development pathways that engage with the city's urban innovation strategies.

2.2 *Beyond Skills: An Inclusive Purpose of Digital HRM*

Inclusion in smart cities covers far more than gender or race; it is about access to digital space, equitable learning, diversity of age, and socioeconomic inclusion [54]. HRM systems that introduce algorithmic decision-making (as compared to traditional decision-making) can provide both opportunity and danger. In one perspective, AI-based tools can reduce bias and widen recruitment funnels [47]; however, poorly designed algorithms could potentially encode existing societal prejudice [38]. Further, digitalized HRM should prove fair, transparent, and inclusive as a guiding principle. Cloud-based systems along with features like multiple language interfaces, remote assessments, and adaptive learning can make sure that those workers that are already marginalized are not excluded from participation in the oligarch digital economy [4]. Ethical frameworks for digital governance and regular content bias audits are essential to maintain the legitimacy of such systems [8].

The gig economy characterizes the urban labor market of today. In smart cities, the gig worker, freelancer and urban micro-entrepreneur populations contribute to innovation, last-mile logistics, and diversity in service offerings. However, these worker populations often fall outside of traditional HRM arrangements. Study recommends that HR ecosystems be expanded in manner to include alternative workforces and run through digital pathways such as smart contracts, mobile dashboards and peer review programs [20]. For example, Blockchain-based Identity systems can provide worker reputation in lieu of formal employee records [45]. COVID-19 has permanently altered work arrangements with remote and hybrid arrangements becoming normalized. In smart cities where the digital infrastructure is strong, this reinforces a move away from centralized workplaces. Nevertheless, remote team management can be difficult to manage in terms of employee engagement, supervision, and well-being [50]. Digital HRM tools can take the form of virtual onboarding platforms to understand an employee's sentiment and how their overall well-being is impacted with the use of digital platforms [22]. Remote workers as described by [9] have improved productivity and job satisfaction when an employee understands expected performance measures and benefits from clear digital collaborative tools.

Big data and people analytics are at the core of the future of HRM in smart cities. The ability to mine workforce data for patterns, predict employee attrition or create individualized learning paths represents a natural evolution into a data-informed field [33]. The ethical norms and frameworks around data protection govern data use. However, HR analytics must be used appropriately. Study states, HR analytics may undermine trust if used to investigate employees or examine data in a punitive way [39]. In smart cities governed by a public–private partnership approach, this concern is exacerbated.

2.3 Frameworks for Human-Centric Digital HRM

Three frameworks provide ways of thinking about how Digital HRM is able to play a role in smart cities:

- *Resource-Based View (RBV)*: Human capital is a VRIN resource, meaning it is valuable, rare, inimitable, and non-substitutable [3]. Digital HRM literature can offer growth opportunities related to VRIN resources through extensive learning, internal mobility, and personalization of experience and engagement. Each of these opportunities provides cities the ability to increase their degree of talent competitiveness [53].
- *Technology Acceptance Model (TAM)*: Technology acceptance is largely determined by perceived usefulness and perceived ease of use [17]. Due to the fact that urban HRM will invariably be hampered by employees that encounter usability issues and/or may not be technologically as adept, factors that guide the adoption of technology will factor significantly into HRM practices of smart cities. Study advanced this model to include social and cognitive variables, which is well-suited for cultural diversity of smart cities [49].
- *Institutional Theory*: Organizations operate under legal, normative, and cultural contexts, and their behaviors are shaped by these contexts. In smart cities, HRM

approaches will have the formidable task of managing changes to legal expectations (such as new privacy regulations); changes in public expectations for HRM (such as employment equity); and compliance due to institutional mandates (such as sustainability) that may justify the HRM approach.

2.4 Proposed Conceptual Model

See Fig. 1.

3 Methodology

This chapter utilizes a conceptual and qualitative methodology that is based on a narrative synthesis of secondary data and a thematic analysis of relevant and pertinent literature, reports and global case practices. An exploratory research design, it aims to investigate how Digital HRM frameworks can be strategically aligned with smart city grand challenges to solve urban workforce issues as they evolve.

Data for this exercise was gathered through a structured review of academic journals, policy documents and institutional reports produced by credible organizations including, the World Economic Forum, McKinsey Global Institute and peer-reviewed HRM and urban studies journals. A literature-based case analysis identified the best practices in selected smart cities (Singapore, Amsterdam and Dubai) and global firms (Google, Microsoft and IBM), to illustrate evidence of innovative Digital HRM implementation.

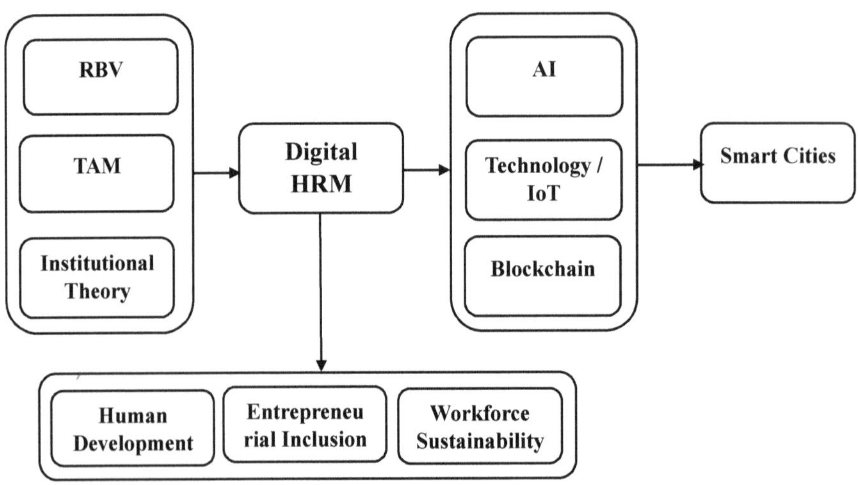

Fig. 1 Proposed conceptual model (*Source* Author)

The analytical insights were finished using a framework that provided a structure with key ideas under the umbrella of: digital HR tools; human-centered design, and sustainable urban workforce. The theoretical framework draws on the theoretical lens of RBV, TAM and Institutional Theory, to propose a conceptual model which identifies connections between digital HRM enablers, entrepreneurial inclusion and workforce resilience to design tech-enabled urban locales.

Although non-empirical in nature, this methodology facilitates critical reflection and conceptual advancement providing a springboard for future empirical inquiry.

4 Role of Digital Transformation in Reshaping HRM Practices Within the Framework of Smart Cities

As cities evolve into smart cities; urban ecosystems characterized by digitally embedded infrastructure and policy; HRM is experiencing an epistemological shift. Digital transformation is not about the development of additional tools, but is fundamentally changing the aims and structures of HRM and calling for a shift in HR practices from transactional functions to strategic, human-oriented functions aligned with urban sustainability, inclusivity, and agility [12, 37].

Artificial Intelligence (AI) is leading the transformation of HR processes to include automated processes in recruitment, learning, and retention. AI-enhanced platforms can now automate screening and interviewing, screen for predictive attrition and engagement insights, and streamline learning opportunities to improve retention [11, 38]. Yet in smart cities mainly knowledge-based work will dominate, and mapping rapidly changing skill requirements to existing and future talent is not straightforward. Thus, predictive analytics can support strategic workforce planning by thinking ahead about capabilities needed to sustain year-on-year innovation and urban competitiveness [27]. This shifts HRM from a supportive space to being part of the key contributors to long-term city planning. Further, **the growing abundance of** urban data made available through the internet of things (IoT) and sensor networks in smart cities provides an opportunity for HRM to become evidence-based and predictive. People analytics enables organizations to examine patterns of behaviour, productivity, and employee health—essential in constructing inclusive and resilient workforces that can rapidly adapt to disruptions [28, 33]. For example, informed decision making by HR managers in smart cities, including equitable distribution of workable human capital or training investments to ensure policy interventions were effective for guiding the optimal human capital strategy of the city.

Smart cities employ platforms to support digital infrastructures, which foster flexible work models and reduce organizations' reliance on central workplaces [32]. All the information about HR [46], such as cloud storage, employee onboarding, and collaboration tools now faithfully demonstrate that digital work is no longer an alternative, but the core structural layer of work [50]. The change in perception and use of this digital infrastructure was exacerbated by the COVID 19 pandemic to the extent it redefined HRM to include dimensions such as digital leadership and employee digital well-being and their ability to steward virtual culture [15]. In this new reality, if HRM can become an active driver of spatial and temporal flexibility, it may simultaneously support organizational objectives and urban sustainability goals through reductions in congestion and carbon emissions based on distributed workforces. Moreover, blockchain technology represents a solution for trust, privacy, and credential verification within smart city ecosystems. Decentralised ledgers could allow verification and sharing of employee data across institutions in a manner that protects privacy and is trustworthy, which can support greater workforce mobility and decrease administrative cost [41]. In cities where there is a high degree of labor fluidity, the type of technologies that bring transparency to HR practices and standard practices around compliance help to enable ethical and inclusive employment practices.

As digitisation of HRM is expanding, so too are the instances of algorithmic bias, employee surveillance, and exclusion. [36] expresses caution regarding narrow organisational definitions of fairness, inclusion and its relation to inclusion, suggesting that digital HRM systems must move toward inclusive design and workplace fairness rather than perpetuating existing inequities. The fundamental reliance on data in smart cities has the potential to devalue autonomy and there is a risk that HRM processes do not contain ethical safeguards. Human-centered design proposes more than just procedural fairness; it considers the balance of efficiency in the use of technologies, individual agency and rights in systems involving high levels of surveillance in the urban environment [40].

5 Workforce Challenges in Tech-Enabled Urban Environments and the Role of Digital HRM in Addressing Them

Smart cities have transformed the socio-economic and technological landscape of city nature. Smart cities, grounded in advanced information and communication technologies (ICT), IoT, AI, and big data, have the potential to promote the quality of life, environmental sustainability, and economic competitiveness [1, 14]. New issues are now emerging, however, as these rapid transformations have raised new, more advanced workforce issues that require a planned change in HRM. Digital HRM (the use of digital tools, platforms, and analytics in managing human capital), will serve to vastly improve and facilitate new HRM processes which will enable organizations to find solutions to these issues as well as prepare a future-ready workforce [13, 30].

5.1 *Major Workforce Challenges in Tech-Enabled Urban Environments*

One of the greatest challenges faced by smart cities is the increasing digital skill gap. The onslaught of AI, automation, cloud computing, and IoT, in particular, has resulted in the demand for a skill base that includes digital literacy, data analytics, cyber security, and agile skills [41]. Yet, research indicates that many employees simply do not have the necessary skills, creating a gap between the competencies of workers and the competencies required in jobs [52]. Further, automation and AI-driven networks in smart cities replace jobs that require routine and manual tasks, and in some cases, even cognitive tasks. This has caused job displacement and is creating an increase in low- and mid- skilled occupations. Simultaneously, there is a growing need for high-skilled, knowledge-intensive jobs; hence the notion of employment polarization. Besides all these, smart cities are creating an innovation ecosystem and are attractive to a globally mobile and diverse workforce. However, as that mobile workforce has more choices, the competition for top talent is intense. Employers are feeling the pressure to attract, engage and retain talent while accommodating increasingly flexible work arrangements, purpose-driven work and well-being of employees [6, 33]. The worker deficit is pushing wages up and turnover across many important sectors of the economy, like technology and health care [24].

Diversity and inclusion are important features of sustainable urban workforces. Navigating generational, cultural, gender, and cognitive diversity in multicultural smart cities is challenging [40]. Excluding inclusive human resource practices from employers' risks disengaging talent, staff engagement, and work performance [36]. Additionally, Smart cities amplify hyper-connectedness, work-life blurring, and high performance. The boom in remote and hybrid work has exposed even more significant risks of burnout, social isolation, and digital fatigue for workers [23, 50]. Supporting mental health and well-being has moved from being an ordinary support to an organizational strategy.

5.2 *How Digital HRM Can Address These Challenges*

Digital HRM uses artificial intelligence (AI) learning management systems (LMS), micro-learning and gamified e-learning to establish clearly focused upskilling and reskilling initiatives as shown in Table 1. Digital learning platforms such as Coursera, Udemy and structured corporate university offerings have begun to find their way into performance management systems, presenting system-based approaches to workforce skill shortages [42]. Predictive analytics displaying skill development needs can also influence interventions [29]. Moreover, data-laden HR analytics can assist organizations in identifying talent needs, find predictors of job loss, or just determine how best to cultivate the workforce [33]. Predictive models can assist HR professionals to prepare for dislocated workers when jobs are automated out of existence or capture when opportunities arise for job rotation or retraining.

Digital HR uses Natural Language Processing (NLP) and machine learning, which aides against recruitment bias, increases ease-of-access to recruitment and increases quality-of-hire through resume screening, candidate matching and simple chatbots that can engage and respond to candidate questions in a timely manner and take on the brunt of FAQs targeting candidates. AI (if employed responsibly) can also discount bias from the recruitment process [11, 38]. Utilization of digital HRM technology can also generate metrics for diversity, sentiment analysis and the possibility of bias-training (through virtual reality (VR) technologies) [40]. Inclusive recruitment platforms and anonymized applications [36] can reduce systemic bias and lead to more equitable hiring practices. Additionally, optimum technology, mental health applications, and tele-counsellors through HR portals are helping organizations monitor and promote the well-being of their employees [22]. Digital HRM has also enhanced opportunities for flexible HR policies and remote/hybrid management by flexibly sharing collaborative technologies to facilitate work-life integration [15]. Moreover, the COVID-19 pandemic sparked digital transformation in HR at a rapid pace. The move to remote work became a normalized method of work. Digital HRM and other platforms like Zoom, Microsoft Teams, Slack and Asana allow for management of a distributed workforce including managing projects and multilateral communication [50]. Performance management also changed, emerging tools allow for tracking objectives and key results (OKR) that could also potentially be edited/updated in real-time.

Table 1 Workforce challenges and solutions

Workforce challenge	Digital HRM solution
Digital skill gaps and technological obsolescence	Digital learning platforms for skill development
Job displacement and employment polarization	Strategic workforce planning with predictive analysis
Talent attraction and retaining talent	AI-enabled talent recruitment and retention
Workplace inclusion and diversity	Advancing DEI
Work-life integration and mental well-being	Promoting well-being utilizing digital well-being devices

6 Real Examples and Best Practices

6.1 Smart Cities Leveraging Digital HRM

Singapore has developed as an international model in using digital HRM in a smart city environment. Digital HRM is evidenced through various Smart Nation and SkillsFuture initiatives, whereby AI-based talent management processes are being adopted across all ministries, to standardize talent sourcing, performance management, and training [35]. The HRP (Human Resource Partnering) transformation framework is also being used to centralize HR services, with an acknowledgment of an increased expectance for strategic planning around workforce management as opposed to HR administration [44]. A distinctive aspect of the country's approach is the personalized upskilling pathways, by way of AI, to provide training recommendations to employees based on their current skills and future role requirements, tailoring learning for engagement of lifetime learning, being adaptable and resilient in employment [51]. This can be noted as keeping a digitally capable public workforce adaptable to new technological advancements.

The importance of including as part of the smart city's digital transformation strategy has been recognized by Amsterdam. In particular, Amsterdam utilizes HRM digital platforms to help facilitate diversity, physical and mental well-being of employees, and hybrid work, amongst a variety of other HR above-mentioned areas. Indeed, Amsterdam uses data-based recruitment tools to promote equity and mitigate biases, as part of the Amsterdam Smart City Platform [48]. Importantly, Amsterdam is also using AI-driven virtual assistants using AI to help employees to connect to HR services, mental health services. City departments also utilize predictive analytics as part of training needs assessments and redeployment of talent to reflect new and emerging urban projects. All of these decisions pinpoint the way to HRM that is people-focused and operates together in a tech-enabled governance environment.

Dubai has implemented blockchain technology in their HR functions, as part of Smart Dubai 2021 initiative, as a method to enable improved efficiency, transparency and automation. The Dubai Electricity and Water Authority (DEWA) is leading the way with blockchain-based HR reminders to verify qualifications, employee records and smart contracts [2]. This provides faster verification with tamper proof credentials providing new hires quicker, streamlined access. Dubai Future Foundation is using AI data-driven predictive tools to map workforce trends and future talent requirements which allows for sound HR planning in line with the emirate's innovation agenda. The continuous data analytical integration around employee engagement and on-off boarding strategies, puts the emirate one step closer to internalizing how all employees can be invested in their own satisfaction in an ever-changing tech-oriented environment [19].

Table 2 Leading smart cities and companies

Company/city	Focus area
Singapore	Development of talent
Amsterdam	DEI initiatives
Dubai	Automation and transparency
Google	Flexibility
Microsoft	Wellbeing
IBM	Talent management (Predictive)

6.2 *HRM Strategies of Leading Tech-Enabled Companies in Urban Settings*

The People Analytics team at Google is one of the best-known examples of digital HRM put into practice. Operated out of the firm's technology hub offices in San Francisco and New York City, Google has developed algorithms that predict when employees are likely to leave, analyze team effectiveness, and determine management potential [10]. Two examples of the approaches Google have taken with data are titled Project Oxygen and Project Aristotle which use HR analytics to improve management effectiveness and team effectiveness specifically. Google has also developed gPeople, a self-service platform that automates elements of HR processes, and creates access to resources about career development. Personalized dashboards that track employee progress and provide nudges for upskilling and wellness interventions work to elevate HR as a strategic partner in employee development as shown in Table 2 [7].

Microsoft's Human Resources (HR) Digital Transformation plan consists of an HR adoption strategy that utilizes AI and machine learning for equitable work environments with urban centers in Seattle and London, Microsoft has introduced AI to identify unconscious bias in job descriptions and performance appraisals [26]. My Analytics and Viva Insights platforms provide users with productivity, collaboration, and well-being feedback based on data—very useful today in the hybrid working outside of a pandemic. Along with their emphasis on using tech, Microsoft's HR model also emphasises on building digital leadership capabilities. Their global LEAP program aims to support transfer into tech roles for underrepresented groups through digital mentoring tools, VR training, and LMS platforms with a focus on inclusivity and retention [16].

IBM is a leader in the use blockchain for HR, to authenticate education and employment credentials in the global market. It is using blockchain in its major city hubs (New York and Bangalore) to reduce hiring cycle times and enable international employee movement. IBM's Your Learning platform provides cloud-based AI personalized learning pathways to facilitate different learning journeys, taking into consideration the goals and feedback of the user. IBM's Watson Talent Insights utilizes big data to identify employee behaviors, flag potential employee turnover issues, and recommend options to encourage engagement. IBM evidence of being digitally agile and creating digital strategic capabilities such as workforce agility efforts, continuing to hire and attract talent in urban locations with volatile labor markets and stiff competition.

7 Discussion

The adoption of digital HRM within smart city frameworks represents a monumental change in the spatial context and strategic deployment of HR practices. While the initial directions of digital HRM primarily increased administrative efficiencies of HR, it has now become a proactive driver of varying degrees of human agency and appetite for new employment dynamics. Through innovative use of technologies in a labor ecosystem, responsible HR now might allow HR to develop sustainable, equitable, and flourishing work environments within socio-technological environments of change. This study suggests that sustaining urban workforces through upcoming socio-ecological challenges means navigation uniquely agile, digitally-enabled HR structures beyond mere efficiencies for socially just workplace environments. The previous literature suggests that digital HRM tools for a local government, such as, AI-mediated recruitment services, technologically savvy digital cloud-based learning systems, and HR analytics, have the potential to make differences in organizational responsiveness and workers adaptability and long-term viability [12, 33]. However, any potential benefits must be framed around the risks inherent in the smart cities' environment, tackling the prospect of job loss, digital disadvantage and, algorithmic discrimination. This study is a timely reminder of the combined risks and rewards provided by digital transformation in HRM: while speed and accuracy of technological tools may provide new and innovative speed and precision, these interventions must be framed to ensure fairness, integrity, and well-being [8, 38].

In addition, smart cities emphasize decentralization and platform work, which might be incompatible with traditional employment. In such a scenario, HRM will need to develop mechanisms to account for the non-standard workers (e.g. gig workers, freelancers, entrepreneurs) who do not have access to traditional employment benefits and learning pathways. Where progress is being made is with additional digital micro-credentialing systems, decentralized identity (blockchain credentials), and ownership records/learning records. In many ways these sets of credentials offer HRM opportunities to support non-standard workers, thereby supporting a more equitable, urban economy [20, 45]. The research also highlights that the successful implementation of digital HRM will be determined by how well it can successfully align to and work with existing institutional and socio-political structures. According to institutional theory, HR innovations are regarded as legitimate with wider acceptance and expectation of society or regulations [37]. The alignment of digital HRM practices must reflect municipalities, citizens, and labor expectations and reflect their aspirations in order to advance practices in the area of HRM in Smart Cities.

The convergence of the multi-dimensional frameworks, namely BRV, Technology Acceptance Model [49] and Institutional Theory, provides a multidimensional view of digital HRM as a strategic enabler of human capital. From this viewpoint, digitally enabled human capital is an important differentiator of urban competitiveness. It is this conversation that confirms that digital HRM, presented holistically, can be the catalyst in creating sustainable, human-centric workforce ecosystems as part of smart cities. Digital HRM must take the responsibility of substantive meaning-making and proceed with a proactive approach based on shaping transitions in the workforce, supporting innovation and anchoring ethical frameworks and commitments to action as cities increasingly digitize.

8 Future Trends and Implications for HRM in Smart Cities

With more cities developing smart technologies, digital HRM will be the third employer in shaping the future workplace through a variety of new technologies that include AI, blockchain, XR, and the metaverse. These technologies promise to bring efficiency, customization, and real-time decision-making for recruitment, training, and engagement [33, 43] to HR practice, but this transformation will also bring challenges. AI may have a tendency to 'over-rely' on software algorithms that could result in algorithmic bias and a decline in human judgment in crucial HR matters [25]. While blockchain has the potential to improvements in transparency and trust through new data storage, its adoption is extremely difficult and does not come without significant issues of regulation and privacy [45]. XR (extended realities, including augmented and virtual realities) and metaverse technologies may facilitate greater participation in the workplace, however they could develop digital fatigue and may come with high infrastructure costs and require excessive workplace time [18]. There will be persistence of themes such as data privacy concerns, surveillance, and concerns about exclusion of digitally disadvantaged groups [8]. Consequently, the future of HRM will have to assess its possibility for shaping smart city workplaces with ethical, inclusive, and sustainable certainties which is quite mandatory. When improving the future of work for HRM, it will be important to strategically align with the organizational ESG (environmental, social, governance) goals and employee wellness programs to ensure smart city workplaces are human-centric, equal, and robust.

9 Limitations and Challenges

Digital HRM in smart cities presents significant opportunities for change; however, it is not without constraints. One of the most critical constraints that influence Digital HRM is the digital divide. Digital divide refers to inequitable access to HR services due to differences in digital infrastructure, as well as differences in employee digital literacy. In addition, with many organizations starting to use AI and analytics, there are privacy and surveillance issues that can impact the overall quality of cultural trust if they are not grounded in ethical practices. Additionally, there are barriers in that some individuals are resistant to adopting technologies, or do not learn about the benefits of advanced digital tools that would allow their immediate, practical use in Work Integrated Learning for HR professionals. Showing increased reliance on technologies may reduce human-to-human interaction and human empathy in times where people need it most to maximize employee engagement experiences. Therefore, overall, while this cannot ignore the efficiency increase of Digital HRM, there is a need to implement it as intentionally as possible with keen regard to inclusion and the commitment to diversification, especially ongoing upskilling in smart cities, for truly inclusive, sustainable and ethically-driven work/power development.

10 Conclusion

The transformation of cities toward smart and technology driven ecosystems provides HRM with strategic opportunities and critical challenges. As urban places become more connected through data, automation, and AI, HRM moves away from purely administrative functions and takes on more of a future-oriented strategic purpose. Digital HRM helps to address significant challenges related to HRM such as skills gaps, algorithmic bias, employee well-being, data privacy, but this also implies HR professionals must be data driven, inclusive, and ethical in their data approaches. Aligning HRM policy and practice with the underlying and embedded goals of smart cities is now more important than ever if organizations want to attract, retain, and develop a skilled, resilient and future-proof workforce. As more and more workspaces become hybrid and digitally mediated, HRM's role in developing inclusive/sustainable work environments is critically important.

References

1. Albino V, Berardi U, Dangelico RM (2015) Smart cities: definitions, dimensions, performance, and initiatives. J Urban Technol 22(1):3–21. https://doi.org/10.1080/10630732.2014.942092
2. Awashreh R, Al Ghunaimi H, Almi H (2025) Leveraging blockchain and smart contracts for advancing sustainability in the GCC's circular economy. Pak J Life Soc Sci 23(1):4221–4234. https://doi.org/10.57239/PJLSS-2025-23.1.00334
3. Barney J (1991) Firm resources and sustained competitive advantage. J Manag 17(1):99–120. https://doi.org/10.1177/014920639101700108
4. Barrett M, Davidson E, Prabhu J, Vargo SL (2015) Service innovation in the digital age. MIS Q. 39(1):135–154. https://doi.org/10.17863/CAM.13034
5. Batty M, Axhausen KW, Giannotti F, Pozdnoukhov A, Bazzani A, Wachowicz M, Portugali Y (2012) Smart cities of the future. Eur Phys J.Spec.Top.214(1):481–518. https://doi.org/10.1140/epjst/e2012-01703-3
6. Bersin J (2017) Catch the wave: the 21st-century career. Deloitte Rev. 21:32–41
7. Bessen JE (2019) AI and jobs: the role of demand. NBER Working Paper No. 24235. https://doi.org/10.3386/w24235
8. Binns R (2018) Fairness in machine learning: lessons from political philosophy. In: Proceedings of the 2018 conference on fairness, accountability and transparency. pp 149–159

9. Bloom N, Liang J, Roberts J, Ying ZJ (2015) Does working from home work? Evidence from a Chinese experiment. Q J Econ 130(1):165–218. https://doi.org/10.1093/qje/qju032
10. Bock L (2015) Work Rules!: insights from inside google that will transform how you live and lead. Twelve, New York
11. Bogen M, Rieke A (2018) Help wanted: An examination of hiring algorithms, equity, and bias. Upturn. https://www.upturn.org/reports/2018/hiring-algorithms/
12. Bondarouk T, Brewster C (2016) Conceptualising the future of HRM and technology research. Int J Hum Resour Manag 27(21):2652–2671. https://doi.org/10.1080/09585192.2016.1232296
13. Bondarouk T, Harms R, Lepak DP (2017) Does e-HRM lead to better HRM service? Int J Hum Resour Manag 28(9):1332–1362. https://doi.org/10.1080/09585192.2015.1118139
14. Caragliu A, Del Bo C, Nijkamp P (2011) Smart cities in Europe. J Urban Technol 18(2):65–82. https://doi.org/10.1080/10630732.2011.601117
15. Choudhury P, Foroughi C, Larson BZ (2021) Work-from-anywhere: the productivity effects of geographic flexibility. Strateg Manag J 42(4):655–683. https://doi.org/10.1002/smj.3251
16. Cooke FL, Dickmann M, Parry E (2023) Building a sustainable ecosystem of human resource management research: reflections and suggestions. Int J Hum Resour Manag 34(3):459–477. https://doi.org/10.1080/09585192.2023.2165011
17. Davis FD (1989) Perceived usefulness, perceived ease of use, and user acceptance of information technology. MIS Q 13(3):319–340. https://doi.org/10.2307/249008
18. Grawitch MJ, Gottschalk M, Munz DC (2006) The path to a healthy workplace: a critical review linking healthy workplace practices, employee well-being, and organizational improvements. Consult. Psychol. J. Pract. Res. 58(3):129–147. https://doi.org/10.1037/1065-9293.58.3.129
19. Hameed Z, Khan IU, Sheikh Z, Islam T, Rasheed MI, Naeem RM (2019) Organizational justice and knowledge sharing behavior: the role of psychological ownership and perceived organizational support. Pers Rev 48(3):748–773. https://doi.org/10.1108/PR-07-2017-0217
20. Jesuthasan R, Boudreau J (2022) Work without jobs: how to reboot your organization's work operating system. MIT Press, Cambridge
21. Kitchin R (2014) The real-time city? Big data and smart urbanism. GeoJournal 79(1):1–14
22. Kniffin KM, Narayanan J, Van Vugt M (2021) COVID-19 is a moderating variable with its own moderating factors. Ind Organ Psychol 14(1–2):149–151
23. Kniffin KM, Narayanan J, Anseel F et al (2021) COVID-19 and the workplace: implications, issues, and insights for future research and action. Am Psychol 76(1):63–77. https://doi.org/10.1037/amp0000716
24. Kraus S, Schiavone F, Pluzhnikova A, Invernizzi AC (2021) Digital transformation in healthcare: analyzing the current state-of-research. J Bus Res 123:557–567. https://doi.org/10.1016/j.jbusres.2020.10.030
25. Leicht-Deobald U, Busch T, Schank C, Weibel A, Scherer A, Claassen C (2019) The challenges of algorithm-based HR decision-making for personal integrity. J Bus Ethics 160:377–392. https://doi.org/10.1007/978-3-031-18794-0_5
26. Lindzo, J (2020) How microsoft uses AI to promote diversity and well-being in the workplace. Fast Company
27. Manyika J, Lund S, Chui M, Bughin J, Woetzel J, Batra P, Sanghvi S (2017) Jobs lost, jobs gained: workforce transitions in a time of automation. McKinsey Global Institute
28. Marler JH, Boudreau JW (2017) An evidence-based review of HR analytics. Int J Hum Resour Manag 28(1):3–26. https://doi.org/10.1080/09585192.2016.1244699
29. Marler JH, Fisher SL (2013) An evidence-based review of e-HRM and strategic human resource management. Hum Resour Manag Rev 23(1):18–36. https://doi.org/10.1016/j.hrmr.2012.06.002
30. Marler JH, Parry E (2016) Human resource management, strategic involvement and e-HRM technology. Int J Hum Resour Manag 27(19):2233–2253. https://doi.org/10.1080/09585192.2015.1091980
31. McKinsey & Company (2018) Jobs lost, jobs gained: Workforce transitions in a time of automation. https://www.mckinsey.com

32. Meijerink JG, Bondarouk T, Lepak DP (2016) Employees as active consumers of HRM: linking employees' HRM competences with their perceptions of HRM service value. Hum Resour Manag 55(2):219–240. https://doi.org/10.1002/hrm.21719
33. Minbaeva D (2018) Building credible human capital analytics for organizational competitive advantage. Hum Resour Manag 57(3):701–713. https://doi.org/10.1002/hrm.21848
34. Ministry of Housing and Urban Affairs (2023) Smart cities mission. Government of India
35. Ng PCD (2018) How do national employment policies influence Human Resource practices? A comparative study of the integrated resort and hotel industry in Macau and Singapore
36. Nishii LH (2013) The benefits of climate for inclusion for gender-diverse groups. Acad Manag J 56(6):1754–1774. https://doi.org/10.5465/amj.2009.0823
37. Parry E, Strohmeier S (2014) HRM in the digital age—digital changes and challenges of the HR profession. Employee Relat 36(4):345–353. https://doi.org/10.1108/ER-03-2014-0032
38. Raghavan M, Barocas S, Kleinberg J, Levy K (2020) Mitigating bias in algorithmic hiring: Evaluating claims and practices. In: Proceedings of the 2020 Conference on fairness, accountability, and transparency, pp 469–481. https://doi.org/10.1145/3351095.3372828
39. Rasmussen T, Ulrich D (2015) Learning from practice: how HR analytics avoids being a management fad. Organ Dyn 44(3):236–242. https://doi.org/10.1016/j.orgdyn.2015.05.008
40. Shore LM, Cleveland JN, Sanchez D (2018) Inclusive workplaces: a review and model. Hum Resour Manag Rev 28(2):176–189. https://doi.org/10.1016/j.hrmr.2017.07.003
41. Sivathanu B, Pillai R (2018) Smart HR 4.0–how industry 4.0 is disrupting HR. Hum Resour Manag Int Dig 26(4):7–11. https://doi.org/10.1108/HRMID-04-2018-0059
42. Stone DL, Deadrick DL, Lukaszewski KM, Johnson R (2015) The influence of technology on the future of human resource management. Hum Resour Manag Rev 25(2):216–231. https://doi.org/10.1016/j.hrmr.2015.01.002
43. Tambe P, Cappelli P, Yakubovich V (2019) Artificial intelligence in human resources management: challenges and a path forward. Calif. Manag. Rev. 61(4):15–42. https://doi.org/10.1177/0008125619867910
44. Tan CL, Nasurdin AM (2011) Human resource management practices and organizational innovation: assessing the mediating role of knowledge management effectiveness. Electron J Knowl Manag 9(2):155–167
45. Tapscott D, Tapscott A (2017) Blockchain revolution: how the technology behind bitcoin and other cryptocurrencies is changing the world. Penguin, New York
46. Tursunbayeva A, Pagliari C, Bunduchi R, Franco M (2017) Human resource information systems in health care: a systematic evidence review. J Am Med Inform Assoc 24(3):633–654. https://doi.org/10.1093/jamia/ocw141
47. Upadhyay AK, Khandelwal K (2018) Applying artificial intelligence: implications for recruitment. Strateg HR Rev 17(5):255–258. https://doi.org/10.1108/SHR-07-2018-0051
48. Van de Klundert J, van Dongen-van den Broek J, Yesuf EM, Vreugdenhil J, Yimer SM (2018) 'We are planning to leave, all of us'—a realist study of mechanisms explaining healthcare employee turnover in rural Ethiopia. Hum Resour Health 16(1):37. https://doi.org/10.1186/s12960-018-0301-0
49. Venkatesh V, Bala H (2008) Technology acceptance model 3 and a research agenda on interventions. Decis Sci 39(2):273–315. https://doi.org/10.1111/j.1540-5915.2008.00192.x
50. Waizenegger L, McKenna B, Cai W, Bendz T (2020) An affordance perspective of team collaboration and enforced working from home during COVID-19. Eur J Inf Syst 29(4):429–442. https://doi.org/10.1080/0960085X.2020.1800417
51. Wong YK (2021) Human resources development using AI. Int J Eng Trends Appl 8(4):56–60
52. World Economic Forum (2020) The future of jobs report 2020. https://www.weforum.org/reports/the-future-of-jobs-report-2020
53. Wright PM, Dunford BB, Snell SA (2001) Human resources and the resource-based view of the firm. J Manag 27(6):701–721. https://doi.org/10.1177/014920630102700607
54. Yigitcanlar T, Kamruzzaman M (2018) Does smart city policy lead to sustainability of cities? Land Use Policy 73:49–58. https://doi.org/10.1016/j.landusepol.2018.01.034

Technology-Enabled Affordable Housing Solutions

Piyal Roy

Abstract Rapid population growth in urban areas, combined with existing social inequalities among populations, has heightened the need for accessible housing. The existing standard building practices demonstrate inadequate performance in terms of affordable and sustainable housing solutions. This chapter studies methods to enhance affordable home selection through current building methods coupled with digital property management systems and intelligent system implementations. The housing sector develops through integrated ventures between Building Information Modeling (BIM) and 3D printing, along with modular construction and Internet of Things (IoT) managed by blockchain-based property management technologies. The union of Artificial Intelligence (AI) technology and data analytics enables the improvement of housing distribution networks, along with design models that align with scholarly recommendations. Both developing and global economies have housing affordability cases serving as the foundation for this study to show how public–private partnerships succeed using fiscal policies and fintech solutions. Technology integration in affordable housing generates financial benefits to boost resource protection approaches while building inclusive practices and establishing higher building quality standards.

Keywords Modular construction · Prefabricated housing · Affordable housing · Architectural design principles · Cost efficiency

P. Roy (✉)
Brainware University, Kolkata, India
e-mail: piyalroy00@gmail.com

V. Chand Sharma et al. (eds.), *Tech-Enabled Urbanism and Entrepreneurship for Inclusive Cities*, Lecture Notes in Civil Engineering 779,
https://doi.org/10.1007/978-981-95-3873-7_4

1 Introduction

1.1 Overview of Affordable Housing Challenges

Accelerated urban population growth, socioeconomic inequality, and rising living standards are all contributors to the ongoing global housing crisis. In most societies of developed as well as developing countries, the demand–supply gap is widening in terms of housing. Particularly in the aftermath of climate change and changing urban trends, traditional construction practices simply fail to meet low-cost and sustainable housing needs. These conditions are compounded by a lack of urban planned integration, bureaucratic congestion, and limited available land, leading to overcrowded cities, slums, and poor infrastructure.

People living in cities can expect economic growth and better living, yet if more people seek housing than housing is available, inequality may grow. Building in the usual way is slow, uses lots of labor and money, and is not suitable for housing on such a large scale. It is hard for them to integrate features that make buildings withstand climate change and support the environment. Also, dealing with complex regulations, long permitting processes, and divided ownership blocks the country's development. We are currently facing an urgent problem because there is a big gap between the number of people seeking housing and the ability of systems to offer equitable, sustainable and inclusive answers.

1.2 Role of Technology in Addressing Housing Affordability

New technologies can solve the different problems of providing affordable housing. Modular building, 3D printing and BIM can make projects quicker, less expensive and use fewer resources compared to older methods. Through these innovations, waste can be minimized, quality control improved and project times can be shortened. At the same time, the IoT, blockchain technology and digital systems used for housing properties increase the security, efficiency and transparency of managing them. AI and data analysis contribute to sound choices in designing and assigning housing, as well as efficient and helpful management of tenants.

Harnessing technology properly and for all people can cause a significant shift in the housing field. With the help of new technology, different approaches and enhanced digital tools, expenses can be reduced, timing is shortened, and the final outcome can be better. For example, the use of modular and prefabricated construction methods lets homes be built in factories, which makes building much faster and generates less waste. Sharing detailed information using BIM allows all parties involved in a project to coordinate and notice issues before construction begins. Just as with electric vehicles, 3D printing can provide homes that cost less to build, require less material and give buyers many personalization options.

Digital technologies also allow for housing that is easier to get, distribute and control. Thanks to IoT, real-time adjustments can be made to energy, security and maintenance, making housing practical and comfortable. Through Blockchain, land records, tenant documents, and ownership verification are more open, which helps fight corruption and fraud. Many cities rely on data processing and AI, which identify crowded places, decide on ideal housing allocation and forecast future changes in demand. These technologies are not entirely independent of the other elements that make up a healthcare system. The fundamental importance becomes clear once these systems are involved in the broader environment of governance, finance and social participation. Adding technology and partnering public and private organizations can improve the equity and accessibility of housing in whole communities.

1.3 Scope and Objectives of the Chapter

This chapter studies how modern technology, such as blockchain, AI, modular buildings, BIM, IoT and 3D printing, can help address the shortage of affordable housing across nations and secure lasting results through proper systems and collaborative efforts. It looks at the common issues linked to shared housing, which exist in both developed and developing countries, and acknowledges that these places have their own unique situations. It discusses how construction can be made more efficient with digital technologies, a better system is designed for housing and property management, and it is made more efficient. The variety of international case studies in the text explains successful uses of technology-assisted housing solutions. Also, it stresses the benefits of innovations in finance, policies that help all and partnerships among the public and private sectors, which help develop housing in a large and fair manner.

2 Technological Innovations in Housing Construction

Innovative construction methods have emerged as pivotal tools for delivering scalable and sustainable housing solutions. Traditional construction techniques are increasingly unsuited to the demands of rapid urbanization, climate resilience, and affordability. In this context, technologies such as modular construction, 3D printing, and Building Information Modeling (BIM) are being implemented globally to accelerate housing delivery while maintaining quality and reducing costs.

2.1 Prefabrication and Modular Homes

Modular and prefabricated techniques have been widely recognized for their efficiency in accelerating construction timelines and reducing costs and waste.

These approaches enable mass production of standardized components in factory-controlled settings, which are then transported and assembled on-site. This reduces environmental impact and improves quality assurance [1].

In countries like the Netherlands and parts of Scandinavia, prefabricated modules are being used to construct energy-efficient housing blocks that meet both aesthetic and functional needs. Hybrid models—combining factory-made components with local labor—are prevalent in India, Kenya, and Brazil [2], addressing employment and material cost concerns simultaneously.

Design adaptability remains critical. For these solutions to be contextually relevant, they must accommodate cultural norms, environmental factors, and spatial dynamics of local communities [3]. Effective implementation thus hinges on thoughtful application of architectural design principles, such as passive ventilation, flexible layouts, and community-oriented spatial planning [4].

2.2 3D Printing for Housing Development

The construction sector is interested in switching from conventional methods of building infrastructure to 3D printing. 3D printing is said to have the potential to shorten construction times and reduce labor and material costs [5]. This will also introduce manufacturing methods and techniques to home building. The backlog of affordable and long-lasting dwellings is constantly expanding on a global scale. Adopting 3D printing for infrastructure has several technological advantages, including more design freedom, accelerated building construction, economic impact, and presumed environmental benefits. The stages involved in the 3D printing process for housing development are outlined in Fig. 1.

3D printing, also known as additive manufacturing, is on the verge of becoming one of the most groundbreaking building technologies of our era. Here, a mechanical machine builds up layers of material, most commonly concrete or composite mixtures, to construct the framework of a house from a digital plan.

The technology also enables complex and curved structural forms that would be expensive and prohibitive to achieve with conventional means [6]. The free-form building enables customized and sensitive home shapes, adapted to unique geographical and climatic requirements.

Organizations in Mexico are working together with 3D printing firms to develop whole neighborhoods for poor families, showing the practical applications of the technology. In the United States and Europe, too, printed houses are being employed as prototypes for low-cost, energy-efficient living spaces. 3D printing homes is a nascent technology, and large-scale deployment involves enormous investment, approval from the regulatory bodies, and skilled manpower to operate and maintain the hardware [7]. Additionally, the long-term durability and environmental viability of the printed houses must be subjected to continuous testing.

Fig. 1 3D printing process for housing development

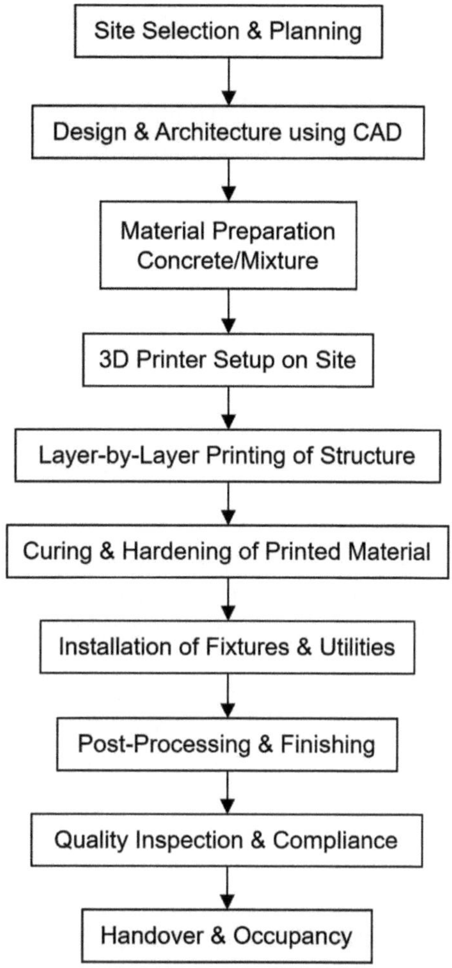

2.3 Sustainable and Smart Materials

Materials that may react in a practical and typically recurring way to changes in their state or the environment they are exposed to are known as innovative materials. We refer to the structures that use smart materials as " Smart Structures". Smart materials are ones that react in a repeatable and predictable manner to changes in the environment, like temperature, light, or pressure. When employed in buildings, they produce smart structures—systems that change in real time to enhance performance and comfort [8]. In low-cost housing, the materials contribute to lower energy consumption and maintenance requirements. Some examples are: Phase Change Materials

(PCMs) for thermal control, thermochromic glass to control natural light, and self-healing concrete to minimize repair requirements. Smart materials enable sustainability by enhancing energy efficiency and increasing building lifespan. Though yet constrained by cost and technical issues, they have great potential for future low-cost, environmentally friendly housing. Solutions [9]. The use of smart and sustainable materials in the construction of houses is another step toward constructing cheap homes that are not only less expensive to build but also more efficient to live in. These materials react to environmental factors such as temperature, light or moisture and improve comfort and use less energy. A case in point is Phase Change Materials (PCMs), which take in and give off heat when the temperature changes, making the use of artificial cooling and heating less necessary. Such an approach is very practical where the temperature fluctuates a lot. Another such example of a fine product is thermochromic glass, which has the capability to respond to sunlight intensity, controlling indoor lighting and cutting down on daytime electricity consumption. Self-healing concrete is another emerging material that can cure its own cracks by chemical reaction, extending building lifespan and reducing the expenditure on maintenance. Even though it costs more than regular concrete up front, long-term cost savings on repair and structure are significant, especially for residences in isolated or economically challenged areas. These products are not only concerning energy—these products also provide occupant health, safety, and well-being [10]. For instance, materials with air-purifying or moisture-sensitivity properties can enhance indoor air quality, which is critical in high-density housing estates. Although the production of such materials is at present limited by cost, continued research and expanding market access are bound to reduce costs through economies of scale. Their use in affordable housing projects needs to be subsidized or incentivized by policymakers and developers as a component of overall sustainability efforts.

3 Digital Platforms and Housing Accessibility

The digitalization of real estate and housing systems has revolutionized how homes are discovered, allocated, and managed—contributing significantly to both transparency and affordability. From PropTech platforms to blockchain-backed registries and AI-driven allocation systems, these technologies are reshaping the housing landscape, especially for low- and middle-income populations.

3.1 PropTech and Digital Marketplaces for Affordable Housing

PropTech (Property Technology) is the application of technology to turn the real estate industry into a digital enterprise, making it easy for users to buy, sell, rent, and manage properties. These technologies have the potential to also help solve issues of affordability by creating more transparent and open housing markets [11].

Online marketplaces offer users the ability to search for housing through filters that represent affordability and, in some cases, incorporate data about location and finance that can be applied to make a choice. Some specifically offer listings for subsidized or low-rent housing, and thus directly serve low-income populations.

PropTech enables virtual property tours, digital contract signing, and an AI chatbot application process to automate the operations, optimizing and minimizing the administrative and overhead costs to bring down the transaction fees between buyers and sellers. Unfortunately, the usefulness of these methods is typically impaired by the digital divide. PropTech refers to technology applications in real estate and property management. It used to focus on high-end transactions but is now being employed to close affordability gaps and create more inclusive housing. Markets [12]. Essentially, PropTech helps people search, apply for, and find housing more efficiently. Internet-based platforms like Zillow, Padmission, or Nest Quest allow users to filter search results based on affordability criteria such as income ranges, rent subsidies, or location. The platforms tend to incorporate built-in financing tools, cost estimators, and spatial data to enable tenants and buyers to make informed decisions.

PropTech also eliminates red tape complexities by introducing features such as electronic lease agreements, virtual tours, and web-based application systems, to the advantage of both housing providers and applicants. For landlords and property managers, automated systems simplify rent collection, maintenance requests, and tenant communications, remain cost-effective, and can be translated into reduced costs.

These advantages notwithstanding, the digital divide—uneven internet and technology access—remains a significant barrier [13]. Most low-income individuals lack access to smartphones, internet subscriptions, or digital literacy skills, and thus are unable to realize the full potential of these innovations. Digital housing websites, therefore, must be supported by outreach campaigns, training, and offline support to enhance universal access.

3.2 *Blockchain for Transparent Housing Transactions*

Blockchain technology is emerging as a transformative tool in housing industry transactions by enhancing trust, transparency, and operational efficiency. As a decentralized and tamper-resistant ledger system, blockchain enables secure authentication of property ownership, transfer of deeds, and lease management without the need for intermediaries.

In the context of affordable housing, blockchain applications address critical challenges such as land title insecurity and transaction fraud. The use of smart contracts—self-executing digital agreements with pre-coded terms—can help ensure that processes such as eligibility verification, financing, and ownership transfer are executed with integrity and transparency [14].

Trust and verification remain central issues in many housing markets, particularly in regions where informal settlements, unauthorized land occupation, and fraudulent title practices are prevalent. Blockchain-based property registers offer a potential solution by creating immutable records of property ownership, leaseholds, and mortgages. These entries are chronologically recorded in digital ledgers that cannot be altered without detection, significantly reducing the risk of corruption or legal disputes [15].

Smart contracts within these systems can be used to automate rental transactions, validate applicant eligibility, and facilitate secure property transfers. Countries such as Georgia, Sweden, and India have piloted blockchain land registry systems with promising results. These initiatives have shown improvements in transaction speed, record security, and public confidence in the real estate sector [15].

3.3 AI and Data Analytics for Housing Allocation

AI and data analysis are playing a growing role in helping us with the allocation and design of housing that is committed to urban planning principles and social fairness. For instance, large datasets on demographic characteristics, income distribution, mobility patterns, and housing supply can be analyzed by AI, and new formulas for affordable housing units could be developed.

Predictive analytics can identify new areas of housing need, model the impact of policy changes, or forecast the trend of long-term affordability measures [16]. AI can also help customize housing design for local needs, making sure developments are not only inclusive but also grounded in context. Also, integration with IoT systems allows building occupation, energy efficiency, and preventive maintenance to be tracked in real-time, leading to reduced operation costs and improved living conditions. However, ethical challenges like algorithmic bias, data privacy, and responsibility must be addressed to ensure equal use of AI in housing [15]. Employed responsibly, data-driven remedies and AI can contribute importantly to the fair and effective allocation of affordable housing resources.

AI and data analytics are transforming the way governments, NGOs, and private developers recognize and address housing needs. Using large data sets—census data and mobility patterns, income levels, and housing supply—AI can identify underserved markets, predict future housing shortages, and locate and design affordable housing projects where they will be most effective. For instance, a predictive analytics application might: Project where housing demand will increase in line with city growth patterns. Model the effect of new policies or zoning regulations. Study the long-term performance of long-term affordability alternatives over the long term. Support fair housing allocation decisions based on actual time needs. Already, in a few cities, AI is being utilized to automate application procedures for public housing. Algorithms can select applicants on a variety of social and economic criteria to ensure that resources go to those in most need. In Singapore and regions of Canada, AI-powered tools are being pilot-tested to optimize social housing allocation by reducing commute time, optimizing proximity to healthcare facilities or schools, and promoting optimal community diversity.

Also, when coupled with IoT (Internet of Things) infrastructure, AI can assist in managing building performance in real time. Smart sensors installed in cost-effective homes can measure occupancy, energy use, water usage, and indoor air quality-directing this data to AI systems that optimize building operations and prevent costly breakages. There are major ethical issues associated with these advantages. If they are not properly managed, algorithms can increase discrimination or fail to notice underrepresented populations. Another concern is the privacy of information, which is especially important with sensitive groups. So, an ethical use of AI requires it to be open, accountable and guided by the community. Digital systems may make it easier and more affordable to construct, keep and use housing. RTIs and other new technologies assist with searchable records online, on blockchain sites and AI analytics, all of which can speed up, balance and improve responsiveness in housing. On the other hand, these systems have to be implemented carefully and with care. Digital innovation might further isolate people if the digital divide remains, data is not well protected, and no effort is made to include communities in designing these tools. Transforming to digital tools should be a way to achieve better, open and fair housing for all and most of all, the vulnerable. Digital solutions can help us create, maintain and give access to reasonably priced housing. Databases linked to searches, records stored on the blockchain and AI helping to plan and arrange housing can make housing systems more effective, just and suited to people's needs [17].

4 Smart Housing and Energy Efficiency

Using smart technology and eco-friendly policies is a major part of modern affordable housing. Due to the increasing number of people moving to cities and the pressure on the environment, affordable housing must be both economical, durable, and environmentally friendly. The chapter explains how IoT technologies, renewable energy solutions, and efficient waste and water management help both the economics and the environment of low-cost housing projects [18]. Urban growth is making it necessary to create housing that stays affordable, is reasonable in costs and is sustainable in future. A bright solution may be to link smart technology and eco-conscious design with housing. In addition to protecting users, smart housing enables living areas that use resources thoughtfully and answer users' needs efficiently. If the IoT and sustainable energy and water systems are used, homes can fit into an economy- and environment-friendly system [19].

4.1 IoT-Enabled Housing Efficiency

IoT technology is revolutionizing living spaces into smart, adaptive places that maximize resource utilization and minimize operational expenses. Smart homes enabled through IoT include integrated sensors, devices, and systems that track and manage utilities like heating, cooling, light, and security in real time. For affordable housing applications, these technologies can cut energy use by as much as possible through adaptive systems, such as smart thermostats that adjust temperature according to occupancy or usage patterns, and lighting systems that shift from electric lighting to natural light depending on availability [20]. Remote monitoring and predictive maintenance also reduce repair costs and downtime.

Smart metering allows residents to monitor their water and energy consumption, which induces conservation habits. For governments and property managers, IoT platforms offer useful information to pinpoint inefficiencies and plan resources efficiently.

Low-income families may see the largest advantage from IoT technologies in terms of cost savings. A smart thermostat will automatically lower heating or cooling when it detects occupancy, the seasons or the time of the day. No energy is used during days when no one is there, and this helps cut down on bills. The same principle works for automated lights, which turn themselves off if the room is not being used and dim their lights during the day to use as little electricity as possible.

It is possible to check and manage utilities in real time and from afar with IoT. It has become simple for people to watch how much water or electricity they use through mobile apps, and this allows them to track down any waste and learn to save. Practices [21].

In some buildings, residents are given the opportunity to view their energy use in comparison to the entire building, which helps them make informed decisions about their living habits. On the maintenance front, smart sensors can pick up on leaks, equipment breakdown, or irregular consumption patterns early, enabling rapid intervention before minor issues turn into expensive repairs. Individual usage dashboards are even offered in certain buildings where tenants can compare their usage with that of the building, leading to more responsible living.

On the maintenance side, smart sensors can detect leaks, equipment failure, or abnormal consumption habits early enough to allow swift intervention before small problems become costly repairs. Predictive maintenance of this kind saves landlords' future maintenance costs and reduces inconvenience for tenants. There are, of course, problems. The initial cost of installing IoT systems—e.g., smart meters, networked appliances, and central platforms—is costly. Data privacy and digital literacy concerns need to be addressed to avoid excluding vulnerable populations from the benefits of smart technology [22].

4.2 Renewable Energy Integration in Affordable Housing

Solar photovoltaic (PV) systems are well-suited for decentralized, cost-effective power generation in both urban and rural affordable housing developments. When integrated with battery storage systems, solar installations can provide a continuous electricity supply and reduce dependence on national grids and fossil fuel-based energy sources [23].

Community solar models where multiple households share access to a centralized solar installation have proven effective in reducing upfront costs and enhancing energy accessibility. PV systems can be deployed on the rooftops of single-family homes or multi-unit buildings to deliver clean, renewable electricity [24]. With battery support, these systems can function during periods of low sunlight, thereby improving reliability and reducing vulnerability to grid instability.

Such models are being implemented in several regions, including the United States, parts of sub-Saharan Africa, and Southeast Asia. These initiatives expand access to clean energy among households that may not have the technical or spatial capacity to install individual systems. Beyond cost savings, renewable energy adoption also enhances energy resilience and contributes to national and international climate mitigation goals [25].

4.3 Water and Waste Management Technologies

Efficient water conservation and waste management are essential elements of sustainable housing. Innovative technologies now make available intelligent, decentralized solutions that meet the needs of affordable housing communities. Water-conserving appliances, greywater treatment systems, and smart sensor real-time usage monitoring all help achieve dramatic levels of water savings and utility bill reduction [26]. Rainwater harvesting systems are especially useful in areas experiencing water scarcity, as a permanent supply for irrigation and toilet flushing. Regarding waste disposal, intelligent bins equipped with fill-level sensors, waste sorting technologies, and compost units reduce landfill reliance and increase recycling. Anaerobic digesters are also incorporated in some housing projects to generate biogas from organic waste, which is used in local energy production. Though the initial cost is high, the long-run savings, health impacts, and environmental gains make the incorporation of such systems in affordable housing schemes justifiable. Beyond energy, water usage, and waste generation are two critical aspects of housing sustainability. In many urban areas, water is either scarce or expensive, and waste disposal systems are overwhelmed. Affordable housing developments can benefit enormously from smart water and waste management technologies that reduce consumption, cut operating costs, and support environmental goals. Smart water meters and leak detection sensors can provide individuals with real-time reports on water consumption so that people and property managers can identify leaks, unusual consumption, or waste [27].

Greywater recycling systems are also a game-changer. Greywater systems treat washing machine, shower, and sink wastewater and reuse it for toilet flushing and irrigation purposes—saving as much as 50% of total water usage. Particularly in drought-ridden regions, greywater systems make affordable homes more independent and eco-friendlier. Rainwater collection, through roof catchment and storage in the ground, provides another option. Rooftop catchment and underground storage are inexpensive and simple to adopt in urban as well as rural settings. The water can be utilized for gardening, cleaning, or toilet flushing, curtailing reliance on municipal sources. On the waste side, intelligent solutions are reducing landfill waste and encouraging recycling. Sensor-equipped waste containers can alert waste collectors when full, facilitating timely collection and minimizing spillage. Some containers have automated sorting systems that sort recyclables, organics, and regular trash, increasing recycling rates and minimizing contamination. Complex systems like anaerobic digesters can potentially transform organic waste (e.g., kitchen scraps) into usable biogas that can be used for heat or cooking. Not only do these types of solutions reduce waste, but they create energy in the local area, closing the loop systems, promoting sustainability and affordability. The cost of these technologies remains an issue for some developments, but in the long run, they conserve water and utilities, extend the lifespan of buildings, and offer healthier environments to live in [28]. Putting intelligent technologies into affordable housing is no longer an extra luxury—it's a must. Climate change, energy deficits, and urbanization are adding up to a set of stresses on the infrastructure, and pioneering housing schemes need to make investment in innovation that supports long-term sustainability and resilience.

5 Financial Innovations in Housing Affordability

Financial innovation contributes significantly to transcending economic impediments to housing access. Conventional financing strategies tend to keep low-income groups out of reach because of credit limitations, expensive interest rates, and bureaucratic processes. To bridge these gaps, various emerging financial technologies and collaborative platforms have emerged to widen access to affordable housing [29]. This section examines the contribution of crowdfunding, microfinance, digital lending platforms, and public–private partnerships to democratizing housing finance. As housing costs outstrip earnings growth and urban land becomes ever more in short supply, financial innovation is becoming the only way to bridge the affordability gap [30].

5.1 *Crowdfunding and Microfinance for Housing*

Crowdfunding sites enable people to pool funds to invest in housing ventures or donate toward a particular housing cause. These platforms—such as Realty Mogul or Fundraise—support efforts in a decentralized mechanism of funding for low-cost housing ventures and enable individual investors to make socially meaningful contributions to schemes. In Third World countries, altruistic crowdfunding has also been used to assist individuals in low-income groups in building or rehabilitating houses. Microfinance, by contrast, provides low-interest, small loans to those with no formal credit history. Organizations such as Grameen Bank and Habitat for Humanity's microfinance initiatives supply funds for home repair, incremental self-construction, or land purchase [31].

Crowdfunding enables ordinary people to come together online and finance a housing project using their pooled resources. Crowdfunding is donation-based (e.g., GoFundMe or Chuffed) or investment-based, with the donors expecting a financial return. In the United States, Realty Mogul and Fundrise allow property investments through small amounts of money, with potential for low-cost housing development and financial inclusion [32].

5.2 *Digital Lending Platforms and Mortgage Tech*

Digital lending platforms and mortgage technologies (referred to commonly as mortgage tech) are transforming the housing finance ecosystem by enhancing speed, access, and individualization in loan processing [33].

Online lenders apply algorithms to gauge borrower risk, authenticate identity and income, and approve loans—sometimes in hours. Such instruments reduce chapter work, lower administrative costs, and enhance access to underbanked communities by considering alternative data to establish creditworthiness.

In affordable housing, online lending platforms enable low- and middle-income families to access housing finance with ease and transparency [34]. Additionally, mobile-first lending solutions are particularly groundbreaking where there is weak banking infrastructure. Their success, however, depends on data privacy, compliance with regulations, and safeguarding against algorithmic bias, which could inadvertently discriminate against marginalized communities.

In emerging markets, digital lenders use mobile-first approaches to reach the unbanked. Others offer micro-mortgages for affordable houses to groups with intermittent earnings who are not reachable through conventional loan products. But they also raise serious concerns. Algorithmic bias can inadvertently discriminate against some groups. Data privacy needs to be protected, especially for vulnerable groups [35]. And there needs to be regulatory adaptation to suit lenders and borrowers alike in the rapidly evolving digital lending ecosystem.

5.3 Government and Private Sector Collaborations

Public–private partnerships involving government agencies and private industry players—banks, developers, and technology companies—will be needed to grow affordable housing finance and fill gaps in the market [36]. Public–private partnerships (PPPs) can channel state incentives combined with private capital to finance large-scale residential projects. Governments can, for instance, provide land concessions, tax relief, or credit guarantees to de-risk private sector investment. The developers, in return, agree to provide housing at discounted prices or include a percentage of affordable housing units in mixed-income developments. Newer models such as land value capture, social impact bonds, and housing vouchers in mobile wallets are being formulated as well. These Help improve fiscal accountability, provide targeted subsidies, and enhance transparency. Singapore, Kenya, and Brazil are examples of how combined financing arrangements can considerably amplify the supply of affordable housing without an adverse impact on financial sustainability. Housing finance innovation is not feasible through anything short of a close partnership between the private and public sectors. Governments provide the ability to provide land, subsidization, and regulation, while private entities provide capital, innovation, and operating capacity.

PPPs are at the heart of this cooperative approach. In a typical PPP, the government may provide land, infrastructure, or tax credits, while private developers commit to building a certain percentage of affordable units. They are incentivized by having lower risk and regulatory support. Singapore and Brazil are among the countries that have successfully adopted PPP models for delivering mass housing. In Kenya, the government has been collaborating with fintech companies to offer digital home ownership options for the low- and middle-income households, such as mobile mortgage apps and savings accounts. The other finance options made available through PPPs include Land value capture, where the developer (in this case, the uplifter) provides a percentage of the uplifted land value (due to improved infrastructure) to an affordable housing fund. Social impact bonds: Private capital is used by investors to fund housing development, and the government repays the investment only if specific outcomes are achieved (for example, a reduction in homelessness). Housing vouchers can be issued electronically and redeemed by qualified tenants or developers through a secure, transparent system. Progressive governments are also backing housing cooperatives and community land trusts, enabling residents to create collective ownership forms with secure tenure and shared financial risk. Transparency and accountability are what are necessary in all these models [37].

Proper planning, however, can unlock mass housing delivery at reasonable costs when supported by enabling legislation and smart fiscal policy. Financial inclusion is essential to housing access. Without the ability to pay for a home, whether through savings, credit, or community support, innovative construction and digital platforms alone will not suffice. The future of affordable housing finance lies in the convergence of technology, policy, and collaboration [38]. Microfinance organizations, crowd-based funding platforms, web-based lenders, and official government authorities all bear a responsibility for the development of a housing market covering all socioeconomic strata.

6 Conclusion and Future Directions

The global affordable housing challenge, shaped by accelerating urbanization, rising income inequality, and environmental vulnerability, necessitates a strategic and integrated application of emerging technologies. This chapter has examined how digital innovations—ranging from PropTech and blockchain to AI, IoT, and 3D printing—are reshaping the planning, construction, and management of affordable housing. These tools have demonstrated the potential to reduce costs, improve construction timelines, enhance transparency in transactions, and optimize housing distribution and maintenance. Financial technologies, including digital lending platforms, microfinance, and crowdfunding, further extend access to credit and capital for underserved populations. However, the realization of these benefits depends on coordinated efforts between governments, private developers, and communities, supported by inclusive regulatory frameworks that promote scalability, equity, and ethical governance.

Future research must address key gaps to ensure that technology-enabled housing solutions are both effective and equitable. There is a need for empirical assessments of the long-term performance of smart homes, including their energy efficiency, cost savings, and user experience. Equally important is the exploration of ethical and policy implications of AI-driven housing allocation and blockchain-based land registries, especially in low-income and informal settlements. Understanding how 3D printing technologies can be adapted across diverse socio-environmental contexts and scaled without exacerbating inequalities remains a pressing area of inquiry. The integration of these technologies into broader housing strategies that align with climate goals, urban planning policies, and social welfare systems will also require interdisciplinary study. Ultimately, technological innovation, when combined with inclusive design and participatory governance, holds promise for transforming affordable housing into a more just and sustainable urban future.

References

1. Nam S, Yoon J, Kim K, Choi B (2020) Optimization of prefabricated components in housing modular construction. Sustainability 12(24):10269. https://doi.org/10.3390/su122410269
2. Djukanovic M, Alegre A, Teixeira Bastos F (2025) Prefabricated solutions for housing: modular architecture and flexible living spaces. Buildings 15(6):862. https://doi.org/10.3390/buildings15060862
3. Adeyemi AB, Ohakawa TC, Okwandu AC, Iwuanyanwu O, Ifechukwu GO (2024) Integrating modular and prefabricated construction techniques in affordable housing: architectural design considerations and benefits. Comprehen Res Rev Sci Technol 2(1):010–019. https://doi.org/10.57219/crrst.2024.2.1.0030

4. Steinhardt DA, Manley K (2016) Adoption of prefabricated housing–the role of country context. Sustain Cities Soc 22:126–135. https://doi.org/10.1016/j.scs.2016.02.008
5. Sakin M, Kiroglu YC (2017) 3D printing of buildings: construction of the sustainable houses of the future by BIM. Energy Procedia 134:702–711. https://doi.org/10.1016/j.egypro.2017.09.562
6. Wilson TT, Mativenga PT, Marnewick AL (2023) Sustainability of 3D printing in infrastructure development. Procedia CIRP 120:195–200. https://doi.org/10.1016/j.procir.2023.08.035
7. de Souza EA, Borges PHR, Stengel T, Nematollahi B, Bos FP (2024) 3D printed sustainable low-cost materials for construction of affordable social housing in Brazil: potential, challenges, and research needs. J Build Eng 87:108985. https://doi.org/10.1016/j.jobe.2024.108985
8. Bhattacharjee J, Roy S (2024) Smart materials for sustainable energy. Nat Resourc Conserv Res 7(1):5536. https://doi.org/10.24294/nrcr.v7i1.5536
9. Tabrizikahou A, Kuczma M, Nowotarski P, Kwiatek M, Javanmardi A (2021) Sustainability of civil structures through the application of smart materials: a review. Materials 14(17):4824. https://doi.org/10.3390/ma14174824
10. Nilimaa J (2023) Smart materials and technologies for sustainable concrete construction. Dev Built Environ 15:100177. https://doi.org/10.1016/j.dibe.2023.100177
11. Rogers D, Maalsen S, Wolifson P, Fields D (2024) Proptech and the private rental sector: new forms of extraction at the intersection of rental properties and platform rentierisation. Urban Stud 61(14):2778–2794. https://doi.org/10.1177/00420980241262916
12. Oladiran O, Dickins L (2024) PropTech and real estate. Innovations. https://doi.org/10.1201/9781003262725
13. Saiz A (2020) Bricks, mortar, and proptech. J Prop Invest Finan 38(4):327–347. https://doi.org/10.1108/jpif-10-2019-0139
14. Nasarre-Aznar S (2018) Collaborative housing and blockchain. Administration 66(2):59–82. https://doi.org/10.2478/admin-2018-0018
15. Yang L, Wang J (2020) Research on real estate transaction platform based on blockchain technology. J Phys Conf Ser 1486(7):072074. https://doi.org/10.1088/1742-6596/1486/7/072074
16. Ali T, Nadeem A, Alzahrani A, Jan S (2020) A transparent and trusted property registration system on permissioned blockchain. In: International conference on advances in the emerging computing technologies (AECT), pp 1–6. https://doi.org/10.1109/aect47998.2020.9194222
17. Uchani Gutierrez OC, Xu G (2022) Blockchain and smart contracts to secure property transactions in smart cities. Appl Sci 13(1):66. https://doi.org/10.3390/app13010066
18. Skeledzija N, Cesic J, Koco E, Bachler V, Vucemilo HN, Dzapo H (2014) Smart home automation system for energy efficient housing. In: 2014 37th International convention on information and communication technology, electronics and microelectronics (MIPRO), pp 166–171. https://doi.org/10.1109/mipro.2014.6859554
19. Reinisch C, Kofler M, Iglesias F, Kastner W (2011) ThinkHome energy efficiency in future smart homes. EURASIP J Embed Syst 2011(1):104617. https://doi.org/10.1155/2011/104617
20. Philip A, Islam SN, Phillips N, Anwar A (2022) Optimum energy management for air conditioners in IoT-enabled smart home. Sensors 22(19):7102. https://doi.org/10.3390/s22197102
21. Hussain B, Hasan QU, Javaid N, Guizani M, Almogren A, Alamri A (2018) An innovative heuristic algorithm for IoT-enabled smart homes for developing countries. IEEE Access 6:15550–15575. https://doi.org/10.1109/access.2018.2809778
22. Ediga P, Mittal A, Rajvanshi S, Habelalmateen MI (2024) Smart energy management: real-time prediction and optimization for IoT-enabled smart homes. Cogent Eng 11(1). https://doi.org/10.1080/23311916.2024.2390674
23. Thadani HL, Go YI (2021) Integration of solar energy into low-cost housing for sustainable development: case study in developing countries. Heliyon 7(12):e08513. https://doi.org/10.1016/j.heliyon.2021.e08513
24. Gil GO, Chowdhury JI, Balta-Ozkan N, Hu Y, Varga L, Hart P (2021) Optimising renewable energy integration in new housing developments with low-carbon technologies. Renew Energy 169:527–540. https://doi.org/10.1016/j.renene.2021.01.059

25. McCabe A, Pojani D, van Groenou AB (2018) The application of renewable energy to social housing: a systematic review. Energy Policy 114:549–557. https://doi.org/10.1016/j.enpol.2017.12.031
26. Bui XT, Chiemchaisri C, Fujioka T, Varjani S (eds.) (2019) Water and wastewater treatment technologies. Springer, Singapore. p 82. https://doi.org/10.1007/978-981-13-3259-3
27. Burkhard R, Deletic A, Craig A (2000) Techniques for water and wastewater management: a review of techniques and their integration in planning. Urban Water 2(3):197–221. https://doi.org/10.1016/s1462-0758(00)00056-x
28. Harremoës P (1997) Integrated water and waste management. Water Sci Technol 35(9):11–20. https://doi.org/10.2166/wst.1997.0327
29. Dokko JK, Keys BJ, Relihan LE (2019) Affordability, financial innovation, and the start of the housing boom. Federal Reserve Bank of Chicago. https://doi.org/10.21033/wp-2019-01
30. Gurran N (2019) Innovations in affordability policies. A research agenda for housing. https://doi.org/10.4337/9781788116510.00022
31. Ashen MJ, Mohammed I, Kunya S, Uthman N, Gambo M (2025). Housing developers' awareness on crowdfunding for housing development project financing in Abuja, Nigeria. African J Environ Sci Renew Energy 17(1):137–158. https://doi.org/10.62154/ajesre.2024.017.010529
32. Luo X, Ge L, Wang C (Alex) (2022) Crowdfunding for microfinance institutions: the new hope? MIS Quart 46(1):373–400. https://doi.org/10.25300/misq/2022/15406
33. Fuster A, Plosser M, Schnabl P, Vickery J (2019) The role of technology in mortgage lending. Rev Finan Stud 32(5):1854–1899. https://doi.org/10.1093/rfs/hhz018
34. Gunin A (2024) The future of mortgage financing in the era of fintech: a systematic review. Futur Econ Law 4(2):31–47. https://doi.org/10.57125/fel.2024.06.25.02
35. Rodima-Taylor D (2025) Land, finance, technology: perspectives on mortgage lending. SSRN Electron J https://doi.org/10.2139/ssrn.5061578
36. Setnikar Cankar S, Petkovsek V (2013) Private and public sector innovation and the importance of cross-sector collaboration. J Appl Bus Res (JABR) 29(6):1597. https://doi.org/10.19030/jabr.v29i6.8197
37. Jayasuriya S, Sandaruwan I, Gajanayake A, Perera B, Kumaraswamy MM (2025) Public-private partnerships in affordable housing: a comprehensive systematic review. https://doi.org/10.2139/ssrn.5244760
38. Davidson NM (2016) Affordable housing and public-private partnerships. https://doi.org/10.4324/9781315565903

Sustainable Urban Planning: Leveraging the Solar Energy Production Process Chain in the Redesign of Allame Majlesi City

Shohreh Ezzatian, Enayat Allah MohagheghNasab, Arash Hajrasooliha, Mohammad Rahimi, and Marzieh Hadi

Abstract This article proposes a robust framework for achieving sustainable urban development through the strategic integration of solar energy production and technology-driven entrepreneurship within urban redesign, defined by the restructuring and improvement of urban spatial networks. Focusing on Allameh Majlesi City as a case study, the research aims to transform the solar energy production process into a catalyst for urban economic development, sustainable income generation, and enhanced quality of life. This framework encompasses the theory of energy transition, technology as an enabler and democratizer of access, and the theory of systemic urban barriers. We demonstrate how the solar energy production process chain can act as an accelerator for urban economic development, sustainable income creation, and improved quality of life. Utilizing data mining techniques, specifically Frequent Pattern Mining and Association Rule Mining, on a sample of 100 stakeholders, we analyze attitudes and preferences regarding solar project development and emerging technologies. The results indicate a strong inclination toward developing technology-driven projects that yield revenue from solar energy production. However, discrepancies were observed among different stakeholder groups regarding the utilization methods, types of urban technological projects, and levels of transparency and participation strategies. Furthermore, the findings, in line with achieving an inclusive city, revealed that the advancement of novel technologies and an increase in stakeholders favoring such development can contribute to the empowerment of low-income populations in Allameh Majlesi City. Ultimately, this study contributes to both theoretical knowledge and practical urban policy by proposing a model for resilient, equitable, and innovative urban ecosystems.

S. Ezzatian (✉) · E. A. MohagheghNasab
Art University of Isfahan, Isfahan, Iran
e-mail: Ezzatian_sh@ut.ac.ir

A. Hajrasooliha · M. Hadi
Tarh_O_Memari Company, Isfahan, Iran

M. Rahimi
Azad University of Isfahan, NajafAbad, Iran

V. Chand Sharma et al. (eds.), *Tech-Enabled Urbanism and Entrepreneurship for Inclusive Cities*, Lecture Notes in Civil Engineering 779,
https://doi.org/10.1007/978-981-95-3873-7_5

Keywords Tech-enabled urbanism · Urban entrepreneurship · Inclusive cities · Equitable urban ecosystems · Sustainable urban planning

1 Introduction

As we navigate the third millennium, cities are undergoing an unprecedented transformation; they function as primary hubs of economic, cultural, and social activities while simultaneously confronting global challenges such as climate change, escalating inequalities, and demographic shifts [1]. Consequently, "sustainable urban management" has evolved from a mere concept into a strategic imperative, integrating various dimensions to preserve urban ecosystems for both current and future populations [2, 3]. Energy, particularly renewable sources like solar power, plays a pivotal role in this transformation. The transition away from reliance on fossil fuels mitigates environmental damage and addresses economic and geopolitical vulnerabilities [4]. The abundance and scalability of solar energy position it as a cornerstone of sustainable urban development, though its effective integration necessitates a comprehensive re-evaluation of the entire energy value chain, including feasibility, production, distribution, consumption, and recycling within urban contexts [5].

Beyond environmental and economic sustainability, urban development must also embrace principles of inclusivity. Developing cities, despite their economic progress, frequently suffer from deep social inequalities and the marginalization of vulnerable groups [6]. The "inclusive cities" paradigm emphasizes democratic access, social justice, and broad participation, which demands technological solutions that empower marginalized populations rather than exacerbate existing disparities [7, 8]. In this regard, technology-driven entrepreneurship emerges as a catalyst for integrating both sustainability and inclusivity. Entrepreneurs leveraging digital technologies—such as the Internet of Things (IoT), Artificial Intelligence (AI), and Big Data—can optimize urban management, enhance energy efficiency, and improve residents' quality of life [9]. However, ensuring that these smart city initiatives genuinely promote social equity and serve all segments of society remains a fundamental challenge [10].

This chapter delves into how these modern paradigms are intertwined, focusing on Allameh Majlesi City as a unique case study within the Iranian context. Located in Mobarakeh County, Isfahan Province, is a new city approximately 30 years old, situated adjacent to steel industries. It possesses significant potential for sustainable development and the integration of advanced technologies, providing a suitable setting for an empirical investigation of the aforementioned challenges and opportunities. With a population of approximately 22,000 inhabitants and extensive state-owned lands (around 11,000 hectares) within its municipal boundaries, the city offers an exceptional opportunity to implement principles of sustainable urban management and optimize large-scale solar energy utilization.

The primary focus of this research is on how technological innovation and entrepreneurship can address inequalities and foster an inclusive and participatory urban environment. Accordingly, the main research question is: "How can technology-driven entrepreneurship facilitate sustainable urban development and enhance the quality of life in Allameh Majlesi City through smart infrastructure?" To address this key question, the following sub-questions are posed: (1) How do energy-centric technologies and smart infrastructure solutions play a role in the urban redesign process? (2) How do technology-based urban entrepreneurship approaches enable equitable access, opportunity creation, and the promotion of social justice within the urban community? and (3) How can technological initiatives and urban redesign democratize the urban ecosystem? By emphasizing the development of resilient, equitable, and innovative urban ecosystems, this research not only enriches the theoretical literature in sustainable and inclusive urban planning but also provides practical guidance for urban planners and policymakers in building flexible, resilient, and just cities in the present and future.

2 Theoretical Foundations

2.1 Energy-Centric Technologies and Smart Infrastructures in Urban Redesign Paradigm

Cities, as primary hubs of innovation and consumption, face significant challenges related to environmental sustainability, climate resilience, and urban quality of life amidst rapid urbanization. The integration of "energy-centric technologies" and "smart infrastructures" is fostering a new paradigm in urban planning, transforming cities from passive energy consumers into active, sustainable ecosystems that prioritize efficiency and equity. This transformation is primarily grounded in the "Energy Transition Theory," which advocates for a shift toward decentralized, renewable-based energy systems, such as rooftop solar panels (distributed renewable energies), as opposed to centralized, fossil-fuel-dependent systems [36, 37]. This transition promotes "energy localism" and the role of citizens as "prosumers," thereby enhancing grid resilience. Supporting technologies include energy storage systems, Building Energy Management Systems (BEMS), Electric Vehicles (EVs) with Vehicle-to-Grid (V2G) capabilities, and smart heating/cooling systems. These components are integrated into "Smart Grids" that facilitate real-time energy monitoring, control, and optimization, aligning consumption with renewable generation and reducing peak loads [38, 39].

"Smart infrastructures" consist of interconnected physical assets integrated with Information and Communication Technologies (ICTs), forming the backbone of "Smart Cities." These infrastructures leverage Big Data from sensors, the Internet of Things (IoT), and Geographical Information Systems (GIS) for real-time analysis of energy consumption, traffic, pollution, and citizen behavior, enabling a transition from reactive to predictive urban planning. This approach advocates for "data-driven urbanism," leading to better decision-making in infrastructure placement, transportation, and energy efficiency. By integrating smart lighting, solar public spaces, EV charging stations, and grid-connected building systems, these technologies guide urban redesign that enhances sustainability, reduces carbon footprints, and improves urban quality of life through access to clean energy and efficient infrastructures.

2.2 Technology-Driven Urban Entrepreneurship and the Opportunity for Social Justice

In today's rapidly urbanizing and technologically driven world, cities serve as crucial epicenters of economic and social innovation. Despite offering immense opportunities, these cities also confront challenges such as escalating inequalities and limited access to resources [1, 11]. To address these issues, "urban entrepreneurship" has emerged as a key strategy for fostering inclusive and sustainable urban development [12]. Specifically, "technology-driven urban entrepreneurship" offers innovative potential to democratize access, expand opportunities, and promote social justice within urban communities. This approach to urban entrepreneurship is rooted in Schumpeter's (1934) theory of "creative destruction," which posits innovation as a catalyst for economic growth. However, modern approaches emphasize the creation of "social value" and "urban value" [13, 14]. These include "social entrepreneurship," aiming to innovatively solve societal problems, and "ecosystem entrepreneurship," which highlights the importance of local networks, institutions, and culture [15].

Within urban contexts, entrepreneurship involves capitalizing on opportunities to improve quality of life, strengthen resilience, and develop dynamic urban ecosystems [16]. Technology startups contribute by developing solutions for smart cities, clean energy, civic digital platforms, and accessibility tools—creating jobs, stimulating local economies, and diversifying economic bases [17]. Digital technologies (e.g., online marketplaces, open-source tools, online learning platforms) lower startup costs and remove barriers, allowing marginalized groups to participate more actively in entrepreneurship. By providing access to information, networking, training, and broader markets, technology empowers these groups, thereby facilitating their involvement in urban decision-making processes [18, 19]. Furthermore, Open Data initiatives enhance transparency and accountability, improve governance, and promote a more equitable distribution of opportunities [20]. Emerging sectors like renewable energy and smart transportation introduce new markets, and the platform economy offers flexible income opportunities, despite challenges such as job security issues [21]. Achieving social justice through technology-driven entrepreneurship necessitates creating inclusive ecosystems that address and eliminate barriers such as the digital divide, limited access to capital, regulatory hurdles, and a lack of support networks [22].

2.3 *Technological Initiatives and Democratic Redesign*

Despite planners' claims of prioritizing public welfare, urban development processes are frequently hampered by structural and systemic barriers that exacerbate inequalities, undermine civic participation, and distort resource allocation [23, 24]. These barriers stem from power structures, discriminatory policies, and institutional rent-seeking behaviors that marginalize vulnerable groups and undermine inclusive urban ecosystems. Technological innovations and innovative urban redesign strategies can act as leverage points to dismantle these barriers and promote urban democratization. Institutional and governance barriers include inefficient bureaucracy, corruption, lack of transparency in decision-making, poor coordination among urban agencies, and reliance on top-down planning [25, 26]. Such issues lead to inequitable resource distribution, development projects that ignore citizens' real needs, and a decrease in public trust in local governments. Theories like "Public Choice Theory" highlight inefficiencies arising from self-serving incentives in the public sector, while "Rent-Seeking Theory" explains how powerful actors influence decision-making to privatize public resources [27, 28].

Economic barriers and spatial inequalities are directly reflected in urban spaces [29]. Spatial segregation based on income and class restricts access to essential services—education, healthcare, transportation—and limits job opportunities and affordable housing for low-income and marginalized groups [30]. These barriers are exacerbated by imbalanced infrastructure investments, policies favoring "rent-seeking urbanism," and land grabbing by powerful actors [31]. Social and participatory barriers include weak channels for civic engagement, public apathy, and the inability of marginalized groups to articulate their needs in urban planning [16]. Power theories demonstrate how dominant groups control agendas and access to decision-making, leaving disenfranchised groups unheard [32]. Consequently, cities become "unjust cities," where resources and opportunities are unevenly distributed [20, 33].

3 Theoretical Framework

This research aims to analyze and guide sustainable development in Allameh Majlesi City by establishing a coherent conceptual framework built upon three key and interconnected theories. These theories are, respectively: (1) The theory of urban sustainability and technological energy transition, (2) The theory of urban entrepreneurship and social value creation through technology, and (3) The theory of inclusive urbanism and the democratization of the urban ecosystem. The selection of this theoretical framework is highly suitable for City due to its specific characteristics, such as high solar energy potential, the availability of extensive state-owned lands, and the need for inclusive socio-economic development. These theories enable us to deeply investigate the city's capacities to transform into a model of sustainable development, improve the quality of life, and achieve social justice through technological innovations and urban redesign.

Allameh Majlesi City, benefiting from intense and continuous solar radiation and access to vast state-owned lands, possesses unparalleled potential for the establishment and development of large-scale solar power plants and distributed energy systems. From the perspective of Energy Transition Theory, this city can become a model for shifting from fossil fuel dependency toward Energy Localism and a Prosumerism role for its citizens [34, 35]. The implementation of smart energy infrastructures in this city will not only reduce energy costs and environmental pollutants but also contribute to enhancing environmental quality and citizens' health, which are key indicators of sustainable development. Therefore, this theory provides a framework for understanding and guiding City's clean energy potentials toward sustainable urban goals. In this vision, technology acts as a catalyst for launching new businesses, increasing transparency and accountability, and ultimately empowering marginalized groups. Decentralized Autonomous Organizations (DAOs) can allow citizens to play an active role as stakeholders and decision-makers in the management and operation of solar projects, sustainably injecting revenues from electricity sales back into the city. This would eliminate traditional administrative and financial barriers, thereby creating tangible social and economic value for the urban community.

Allameh Majlesi City, as a relatively new and developing city, presents a unique opportunity to implement the principles of inclusive and democratic urbanism from its inception. This theory helps us identify potential or existing barriers to the equitable distribution of opportunities and resources in this city. The approach of democratizing the urban ecosystem through technological innovations, especially the use of DAOs, can engage citizens in the management of solar projects and related decision-making. This participation not only increases transparency but also ensures the equitable distribution of benefits derived from solar energy development, directly contributing to the realization of spatial and social justice in Allameh Majlesi City. This framework provides a foundation for creating equal opportunities for collaboration among various groups within the city and for developing human-centric smart infrastructures. Given its existing unemployment rate and need for economic development, this City can greatly benefit from this theory. Investment in the solar power industry and the development of related technologies will create opportunities for launching startups in manufacturing, services, maintenance, and also smart entrepreneurship. This will directly contribute to reducing the unemployment rate and increasing sustainable incomes for citizens. The idea of implementing self-governing participatory structures, such as Decentralized Autonomous Organizations (DAOs), gains significant meaning within this theoretical framework. (See Fig. 1).

As depicted in Fig. 1, the process chain of electricity generation from solar energy is proposed as the focus of technology-driven entrepreneurship in Allameh Majlesi City. Through electricity generation, there is potential for sales, exports, and wealth creation within the city. This wealth creation will generate sustainable income for the city, especially if a self-governing associative structure (DAOs) is realized in the management process. Such a structure would allow citizens to act as shareholders

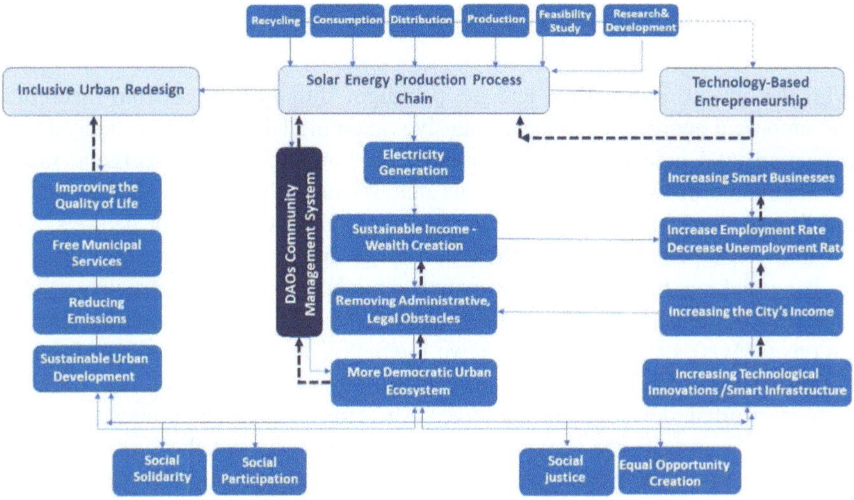

Fig. 1 The effects of the solar energy production process chain on fostering tech-driven urban entrepreneurship and the redesign of inclusive urban infrastructure and spaces

and decision-makers at the management and decision-making levels. This would eliminate traditional administrative, financial, and legal barriers to collaboration in urban entrepreneurship, and is expected to foster a more sustainable urban ecosystem within the city. The effects of creating such a management and economic cycle in the city will, on one hand, lead to the development of urban entrepreneurship across various industrial and service infrastructures.

On the other hand, it will provide opportunities for urban redesign to create an inclusive city. The first branch brings technology-driven entrepreneurship to the city, combining simple manufacturing, service, and maintenance businesses with innovative and smart ventures, leading to the growth of technology and high-tech economies. From the outcome of this entrepreneurship, an increase in employment rates and a decrease in unemployment in the city can be expected. The second branch analyzes the impacts of the solar energy generation process chain on the design of inclusive urban spaces in Allameh Majlesi City. Enhancing the quality of life, the possibility of providing free services, and reducing pollution due to an increase in electric vehicles and smart structures reliant on renewable energy can ultimately lead to sustainable urban development. If the urban ecosystem becomes more democratic and smart infrastructures and technological initiatives are utilized for sustainable urban development, it can be expected that social justice and equal opportunities for collaboration among different groups in the city will emerge.

4 Methodology

With the advent of Big Data and the paradigm shift in science from theoretical and empirical approaches to Data-intensive Scientific Discovery, data analysis and collection, and indeed the entire research methodology, have undergone fundamental transformations. Today, Big Data analysis methods, or "Data Mining," enable the discovery of hidden patterns, relationships, and trends within datasets. This capability has become a competitive advantage and a powerful tool for strategic decision-making. Integrating these analytical methods with novel participatory data collection approaches, such as surveys within Decentralized Autonomous Organizations (DAOs), can create an effective interactive and participatory platform for the management of urban projects, especially when combined with crowdfunding to support urban development initiatives. Therefore, this research aims to identify patterns derived from the responses to a questionnaire by stakeholders of the solar power plant project in Allameh Majlesi City, by combining participatory data collection methods (questionnaire) and data-driven analytical methods (Frequent Pattern Mining and Association Rule Mining). The use of these methods can help in diagnosing the tendencies and perceptions of these project stakeholders regarding novel technologies, participation mechanisms, and investment decisions.

Frequent Pattern Mining and Association Rule Mining techniques were selected due to their ability to identify hidden structures within categorized survey data. Unlike inferential statistics, which typically examine the effect of one variable on another, pattern-based techniques allow researchers to comprehensively observe simultaneous and complex relationships among multiple variables. For instance, patterns such as: "Individuals aged 30–40 with a Master's degree or higher and an inclination to invest, are likely to prefer governance models based on transparency" can be identified. This capability makes these methods highly suitable for examining multi-factorial behavioral patterns in the context of technology adoption in urban environments.

Frequent Patterns and Association Rules: Frequent patterns are sets of items or events that appear repeatedly in a dataset, offering insights into implicit relationships. For example, in customer purchase data, a pattern like "buying milk and bread together" is a frequent pattern. In this study, each response to the questionnaire is considered an item, and combinations of responses form an Itemset. To determine which patterns are frequent, a metric called "Support" is used. For two items X and Y, support is the ratio of the number of transactions containing both X and Y to the total number of transactions. Therefore, high support indicates the commonality of a rule. The FP-Growth algorithm was used for discovering these patterns due to its high efficiency and lack of need for generating frequent candidate item sets.

After discovering frequent patterns, Association Rules are used for simpler interpretation of the results. In other words, association rules are extracted from frequent patterns and express causal or correlational relationships between items in an "If X then Y" ($X \Rightarrow Y$) format. In this rule, X is known as the Antecedent and Y as the Consequent. For example, the rule "If a customer buys milk (X), then they are likely to also buy bread (Y)" is an association rule. The "Confidence" metric is used to extract association rules. Confidence indicates the probability of the Consequent (Y) occurring given the occurrence of the Antecedent (X).

4.1 Research Method

Research Approach: This research employs a quantitative and exploratory approach, utilizing data mining techniques, including Frequent Pattern Mining and Association Rule Mining, to identify hidden patterns in stakeholder responses. The objective of this approach is to provide evidence-based insights for technology-driven and inclusive urban development.

Population and Sample: The statistical population of this study comprises all stakeholders of the Allameh Majlesi City solar power plant project, including residents, investors, technical experts, and institutional representatives. A sample of 100 individuals from this population was selected randomly. Given resource constraints and the exploratory nature of the research, this study is defined as a preliminary (pilot) study, with the possibility of increasing the sample size in future stages, particularly if a DAO is implemented.

Data Collection: Data were collected using an online questionnaire. This questionnaire consisted of 10 closed-ended and categorized questions covering areas such as education, gender, role in the project, willingness to invest, profit-sharing methods, level of desire for transparency, meeting frequencies, decision-making methods, willingness to develop novel technology-based projects from project profits, suggestions for future projects and their types, subsidized and discount measures for residents regarding energy, and other topics related to the City solar power plant. This questionnaire was completed online by citizens of Allameh Majlesi City (See Fig. 2).

5 Founding

In data analysis, Frequent Pattern Mining and Association Rule Mining methods were employed. Generally, frequent pattern discovery methods are categorized into two types: top-down methods and growth-based methods. The FP-Growth algorithm is the most well-known growth-based method for discovering frequent patterns. These methods utilize the "support" metric to determine pattern frequency. In this study, the support value was set to 0.1, and the minimum number of items in each itemset was set to 2.

To execute the FP-Growth algorithm on the data, initial data preprocessing was performed. During the preprocessing phase, some numerical variables, such as age, were discretized into specific age ranges using a categorization method to make them more suitable for the data mining algorithm. Subsequently, in the second stage (data

Fig. 2 Location of the New City of Allameh Majlesi, Isfahan Province, Iran

processing), modeling was carried out, and 193 frequent patterns were extracted. In the third stage (post-processing), "interestingness measures" were used to reduce the number of patterns and highlight more significant ones for visualization and final presentation. Interestingness measures are divided into two categories: objective measures and subjective measures. Here, subjective measures were used. These criteria, based on user input and considerations such as meaningfulness, novelty, practicality, and simplicity, lead to the identification of more "interesting" patterns. In these methods, the user leverages prior knowledge to eliminate repetitive or obvious patterns and select newer, more surprising, or more practical patterns. Accordingly, after refining the patterns from the 193 obtained, the following were selected as the most interesting patterns (See Table 1).

In Table 1, "Items" refers to the combination of options selected by respondents, which have been recorded in a categorized format. "Size" indicates the number of items in each itemset. The "Score" column represents the pattern's "interestingness score" derived from an objective method, which is a composite of Support, Conviction, and Lift values, calculated and provided by RapidMiner software, serving as an auxiliary indicator for the relative prioritization of patterns in the output. After identifying the selected patterns, "If–Then" rules were extracted using the association algorithm and the Confidence metric. The confidence threshold in this research was set to 0.8. This value was chosen based on similar studies and with the aim of focusing on impactful and stable patterns.

Table 1 Selected frequent patterns that can answer the research questions

Items	Size	Support	Score
Focus on current technologies, financial aid for low-income families	2.0	0.3	1.4
Focus on current technologies, mid-season	2.0	0.3	1.2
Focus on current technologies, Bachelor's degree	2.0	0.2	1.0
Focus on novel technologies, highly transparent (full reports, public meetings)	2.0	0.2	1.7
Focus on novel technologies, monthly	2.0	0.2	1.4
Focus on novel technologies, Master's degree and higher	2.0	0.2	1.3
Focus on current technologies; financial aid for low-income families; moderately transparent	3.0	0.1	1.6
Focus on current technologies, financial aid for low-income families, reallocate budget to another city needs	3.0	0.1	2.9
Focus on current technologies; financial aid for low-income families; reallocate budget to another city needs	3.0	0.1	2.0
Focus on current technologies, mid-season, no, I prefer to follow as an audience only	3.0	0.1	2.0
Focus on innovative technologies, Master's degree and higher, monthly	3.0	0.1	2.1
Focus on current technologies; financial aid for low-income families; Smart grid management and urban smarting	3.0	0.1	1.4
Focus on current technologies; financial aid for low-income families; Female; Reallocate budget to another city needs	4.0	0.1	5.5
Focus on current technologies; financial aid for low-income families; Equipping streets and roads with mobile charging technologies	3.0	0.1	1.9
Focus on innovative technologies, very transparent (full reports, public meetings), replacing gasoline cars with electric cars, and building fast-charging stations across the city	3.0	0.1	3.4
Focus on novel technologies; Bachelor's degree; Yes, completely free or fully subsidized	3.0	0.1	2.0
Focus on novel technologies; Master's degree and above; Replacement with new and advanced panels	3.0	0.1	2.2
Focus on current technologies; moderately transparent; Replacement with new and advanced panels	3.0	0.1	1.8
Focus on novel technologies; citizen/Local resident; Highly transparent (full reports, public meetings)	3.0	0.1	2.5
Focus on novel technologies; Bachelor's degree; Highly transparent (full reports, public meetings)	3.0	0.1	1.9
Focus on novel technologies; Master's degree and above; Highly transparent (full reports, public meetings)	3.0	0.1	2.2
Focus on current technologies; moderately transparent; public awareness and education	3.0	0.1	1.8
Male; Focus on current technologies; no, market prices should be respected	3.0	0.1	1.1

(continued)

Table 1 (continued)

Items	Size	Support	Score
Sometimes, when invited; Master's degree and above; Monthly	3.0	0.1	1.6
Focus on current technologies; mid-season; Master's degree and above	3.0	0.1	1.3

Interpretation of Results: An exploration of the derived association rules and an attempt to combine and conclude from these rules led us to interesting findings. These results can contribute to the formation of DAOs, the method of share distribution and management of the solar power plant project, and also the discovery of the relationship between technology-driven entrepreneurship with sustainable development, quality of life, and social justice. The combined interpretations of the association rules are as follows:

- Stakeholders who believe that the focus should be more on existing technologies advocate for aid to low-income families. In contrast, stakeholders who believe that the focus should be more on novel technologies advocate for greater assistance, proposing that energy services for these groups be entirely free or fully subsidized through project profits.
- Stakeholders who believe that the focus should be more on novel technologies possess a higher level of education compared to stakeholders who believe that the focus should be more on existing technologies.
- Stakeholders who believe that the focus should be more on novel technologies, compared to stakeholders who believe that the focus should be more on existing technologies, show a greater willingness for more frequent and shorter-interval participation in the project's steering committee and management.
- Stakeholders who believe that the focus should be more on novel technologies, compared to stakeholders who believe that the focus should be more on existing technologies, demand greater transparency and the provision of full reports and public meetings regarding project management.
- Stakeholders who believe that the focus should be more on existing technologies advocate for reallocating the budget to other city needs after the end of the panels' lifecycle. In contrast, stakeholders with a belief in focusing on novel technologies prefer that worn-out panels be replaced with newer and more advanced panels.
- Stakeholders who believe that the focus should be more on existing technologies, for smart city development and improving quality of life, advocate for equipping streets and roads with mobile charging technologies. In contrast, stakeholders with a belief in focusing on novel technologies prefer that gasoline cars be replaced with electric vehicles and the construction of fast-charging stations across the city be initiated from the profit generated by the solar power plant's electricity sales.

Interpretation of Findings in Response to Research Questions:

The discovered frequent patterns and association rules concerning the relationships between energy-centric technologies and smart infrastructure solutions indicate that the higher individuals' education levels, the more inclined they are toward novel and smart technologies. These individuals prefer that the income generated from energy-centric technologies be spent on novel technologies, smart city development, and the replacement of gasoline vehicles with electric ones. They also wish for solar panels to be replaced with newer and more advanced panels once they reach the end of their lifecycle, and for the solar power plant's operations to be expanded. Therefore, it appears that a higher proportion of stakeholders with advanced education (Master's degree and above) significantly impacts the sustainability and development of the solar power plant, and the widespread adoption of novel and smart city-related technologies.

The discovered frequent patterns and association rules regarding how equitable access, opportunity creation, and social justice can be achieved in the community through urban entrepreneurship approaches reveal that individuals who prefer the development of novel technologies, instead of using existing technologies in the city, have a greater willingness to provide services (in this case, electricity) completely free or with full subsidies for low-income populations. This indicates that an increase in stakeholders with an inclination toward novel technologies can contribute to enhanced social justice in Allameh Majlesi New City.

The discovered frequent patterns and association rules concerning the democratization of the urban ecosystem through technological initiatives show that individuals who believe in the development of novel technologies are much more participatory, are willing to participate and hold meetings at shorter intervals (monthly), and are prepared to take on roles in project management and the steering committee. They also demand maximum transparency in project management, deeming full reports and public meetings essential. Therefore, it seems that efforts to increase the number of stakeholders inclined toward novel technologies can itself be a factor in democratizing the urban ecosystem and enhancing its transparency.

6 Discussion

In Allameh Majlesi City, harnessing natural potentials, especially solar energy, can provide significant opportunities for urban economic development processes and improved quality of life. Specifically, investment in the solar power industry plays a crucial role in creating direct and indirect employment, increasing sustainable incomes, and developing urban infrastructure. Based on the city's demographic and economic development data, the current population is approximately 22,000, and the unemployment rate stands at 13%. With the development of this industry, it is predicted that employment in technical, production, and service sectors related to renewable energies will increase, consequently reducing the unemployment rate. In addition, the city's revenues from the sale of solar-generated electricity can be a stable and substantial source of income, given the region's potential for intense and continuous solar radiation, which facilitates energy sales and revenue generation. The Allameh Majlesi City's provincial and regional electricity distribution grid, due to its proximity and adequate capacity, significantly streamlines the process of distributing and selling generated electricity. Furthermore, the region possesses vast state-owned lands that can be utilized for the construction of large-scale solar power plants. This advantage provides reduced initial investment costs, simplified permitting and administrative processes, and full control over strategic project development.

The preceding sections meticulously laid out a comprehensive theoretical framework and a robust methodological approach for fostering sustainable urban development in Allameh Majlesi City through the strategic integration of solar energy production, technology-driven entrepreneurship, and inclusive urban planning. The empirical findings, derived from the application of Frequent Pattern Mining and Association Rule Mining to stakeholder perspectives in City, provide valuable insights that warrant a nuanced and localized discussion. This section interprets the findings in light of the theoretical framework, emphasizing their implications for urban management and sustainable development within the unique context of the City.

Reaffirming Framework Integration and Stakeholder Empowerment in Allameh Majlesi: The study's results strongly reaffirm the intricate interdependencies among the environmental, economic, social, and governance dimensions of our theoretical framework, which is increasingly recognized as fundamental for effective urban management. The observed patterns, particularly within Allameh Majlesi City, indicate a significant co-occurrence between preferences for transparent governance and a willingness to invest in advanced technologies. Specifically, the findings reveal that stakeholders with higher education (Master's degree and above) and a preference for focusing on novel technologies not only show a greater inclination toward smart urban solutions (e.g., "replacing gasoline vehicles with electric vehicles and building fast charging stations across the city") but simultaneously demand maximum transparency (complete reports, public meetings) and active participation in project management (desire for monthly meetings).

This finding is crucial for City, as it suggests that trust, fostered through transparent and participatory processes, can directly enable the adoption of innovative and sustainable solutions. This goes beyond a simplistic technological determinism and instead affirms the necessity of human-centered approaches and robust participatory governance in the deployment of urban technologies, especially in projects like the solar power plant. Indeed, the data indicate that key stakeholders in Allameh Majlesi, who can be primary drivers of energy transition and smart development, themselves advocate for governance frameworks that hold the potential to democratize the urban ecosystem.

Elucidating Stakeholder Preferences and Advancing Social Equity in Allameh Majlesi: While there is clear enthusiasm for technological advancement and sustainable practices among stakeholders, the data also reveal critical divergences in priorities that demand careful consideration. The recurring pattern indicating a preference to "focus on current technologies" alongside "financial aid for low-income families" and "reallocating the budget to other city needs" signifies a pragmatic and socially conscious perspective. This does not necessarily denote a rejection of innovation, but rather an emphasis on immediate human needs and equitable resource distribution, aligning with key findings in Spatial Justice theory and the concept of "just cities".

This finding is profoundly significant for urban planners in Allameh Majlesi, reminding us that the pursuit of "smart" urban development must not overshadow the imperative of building "just" cities. The core challenge in this city lies in harmonizing top-down technological ambitions with bottom-up community needs and priorities, particularly for marginalized groups who often bear the brunt of developmental inequalities. Interestingly, the findings also demonstrate that even those stakeholders who favor novel technologies are more willing to offer services (such as electricity) entirely free or heavily subsidized for low-income groups.

The Transformative Potential of Decentralized Autonomous Organizations (DAOs) in Allameh Majlesi: Although a DAO was not practically implemented in the data collection phase of this research due to limitations, the conceptualization of Decentralized Autonomous Organizations (DAOs) as a self-governing cooperative structure for managing solar energy projects represents a particularly innovative aspect of this study. The potential for DAOs to democratize investment, empower citizens as decision-makers, and remove traditional bureaucratic and financial barriers is substantial, leveraging blockchain technology's inherent characteristics of transparency and immutability.

The findings of this study strongly support this potential. The discovered patterns indicate that individuals who believe in developing novel technologies in the City are significantly more participatory, tend to organize meetings more frequently (monthly), and are willing to take active roles in project management. Furthermore, they demand maximum transparency in project management, deeming complete reports and the holding of public meetings essential. These results point to the existence of a strong base of stakeholders in the city who align with the fundamental principles of DAOs (participation, transparency, self-governance). Therefore, Allameh Majlesi City can serve as a "living laboratory" for testing these principles, and the success of DAOs here could offer a compelling blueprint for other emerging urban centers seeking to navigate the complex challenges of sustainable development and achieve active and informed citizen participation in an increasingly technology-driven world.

7 Conclusion

This research provides compelling evidence that sustainable urban development in the twenty-first century hinges upon a synergistic integration of advanced energy technologies, inclusive entrepreneurship, and the democratic redesign of urban ecosystems. Through a meticulous examination of the City, and by empirically analyzing the behavioral patterns and preferences of its stakeholders, this study has not only theoretically underpinned this comprehensive approach but has also clearly demonstrated the critical interplay of these factors in shaping community perceptions and the potential for a more equitable and resilient urban future. Our main findings from Allameh Majlesi City highlight that while there is a general appreciation for technology's role in urban sustainability, its application must be tempered with a profound commitment to social justice and the direct alleviation of community needs. This was particularly evident in the contrast between groups focused on existing technologies (prioritizing direct aid to low-income segments and budget reallocation) and groups interested in novel technologies (inclined toward offering free or fully subsidized services), underscoring the necessity for a holistic and inclusive approach in urban planning.

The innovative proposal of Decentralized Autonomous Organizations (DAOs) for managing solar energy projects within this context represents a paradigm shift toward decentralized governance. The willingness of innovation-oriented stakeholders to engage in active and transparent participation, repeatedly revealed in this study's findings, reinforces the idea that DAOs can offer a tangible mechanism to dismantle traditional systemic barriers, foster genuine co-ownership of urban assets, and strengthen democracy within Allameh Majlesi's urban ecosystem.

References

1. UN-Habitat (2020) World cities report 2020: the value of sustainable urbanization. UN-Habitat, https://doi.org/10.18356/b7100b1b-en
2. Jabareen Y (2006) A new conceptual framework for sustainable urban development. Environ Plann B Plann Des 33(1):22–42. https://doi.org/10.1068/b32047
3. Cities Alliance (2018) Cities alliance inclusive cities: the case for a new urban agenda
4. Smil V (2017) Energy and civilization: a history. MIT Press. https://doi.org/10.7551/mitpress/9780262035774.001.0001
5. Sovacool BK (2023) Energy policy: strategies for a sustainable future. Routledge. https://doi.org/10.4324/9780203117248
6. Roy A (2011) Slumdog cities: rethinking urban poverty. Int J Urban Reg Res 35(2):223–238. https://doi.org/10.1111/j.1468-2427.2011.01053.x
7. Sassen S (2013) Cities in a world economy, 4th edn. Sage Publications
8. Batty M (2013) The new science of cities. MIT Press
9. Kitchin R (2014) The real-time city? Big data and smart urbanism. GeoJournal 79(1):1–14. https://doi.org/10.1007/s10708-013-9516-8
10. Shelton T, Zook M, Graham S (2015) The politics of urban infrastructure and the "Smart City" Agenda. Environ Plan A 47(3):515–531. https://doi.org/10.1068/a140050p
11. Florida R (2017) The new urban crisis: how our cities are increasing inequality, segregation, and the failing of the middle class. Basic Books
12. Audretsch DB, Lehmann EE (2016) Urban entrepreneurship and innovation ecosystems: from the entrepreneurial city to the innovative city. Innov J Soc Sci 29(2):119–132
13. Kropp M, Zelepos D (2011) Entrepreneurship and urban development: an analysis of European cities. Springer. https://doi.org/10.1007/978-3-642-15904-4
14. Lerner J (2010) Boulevard of broken dreams: why public efforts to boost entrepreneurship and venture capital have failed–and what to do about it. Princeton University Press. https://doi.org/10.1515/9781400834375
15. Stam E (2015) Entrepreneurial ecosystems and regional policy: a sympathetic critique. Eur Plan Stud 23(9):1759–1768. https://doi.org/10.1080/09654313.2015.1054246
16. Arnstein SR (1969) A ladder of citizen participation. J Am Inst Plann 35(4):216–224. https://doi.org/10.1080/01944396908677247

17. Florida R (2002) The rise of the creative class: and how it's transforming work, leisure, community and everyday life. Basic Books. https://doi.org/10.2307/3552294
18. Sen AK (1999) Development as freedom. Alfred A. Knopf
19. Shirky C (2008) Here comes everybody: the power of organizing without organizations. Penguin Press
20. Lathrop D, Ruma L (eds.) (2010) Open government: transparency, collaboration and participation in practice. O'Reilly Media, Inc.
21. Srnicek N (2017) Platform Capitalism. Polity Press
22. Spigel B (2017) The relational organization of entrepreneurial ecosystems. Entrep Theory Pract 41(1):49–72. https://doi.org/10.1111/etp.12177
23. Harvey D (2009) Social justice and the city. University of Georgia Press. https://doi.org/10.1002/9781444315216
24. Sassen S (2005) The global city: Introducing a concept. The global cities reader, pp 160–167
25. Pierre J (ed) The oxford handbook of urban politics. Oxford University Press. https://doi.org/10.1093/oxfordhb/9780199547530.001.0001
26. Flyvbjerg B (1998) Rationality and power: democracy in practice. University of Chicago Press, Chicago. https://doi.org/10.2307/3005579
27. Tullock G (1967) The welfare costs of tariffs, monopolies, and theft. Western Econ J 5(3):224–232. https://doi.org/10.1111/j.1465-7295.1967.tb01160.x
28. Krueger AO (1974) The political economy of the rent-seeking society. Am Econ Rev 64(3):291–303. https://doi.org/10.2307/1807788
29. Soja EW (2010) Seeking spatial justice. University of Minnesota Press. https://doi.org/10.5749/minnesota/9780816666672.001.0001
30. Marcuse P, van Kempen R (eds) (2000) Globalizing cities: a new spatial order? Blackwell Publishers. https://doi.org/10.1002/9780470712887
31. Sassen S (2014) Expulsions: brutality and complexity in the global economy. Harvard University Press. https://doi.org/10.2307/j.ctt6wprh7
32. Foucault M (1980) Power/knowledge: selected interviews and other writings, 1972–1977.–1977. Pantheon Books. (1980).
33. O'Reilly T (2011) Government as a platform. Innov Technol Govern Global 6(1):13–42. https://doi.org/10.1162/INOV_a_00056
34. Macintosh A (2004) Characterizing e-participation in policy-making. In: Proceedings of the 37th Annual Hawaii International Conference on System sciences, 2004. IEE, p 10. https://doi.org/10.1109/HICSS.2004.1265275
35. Sloane M (2017) Smart cities and the new urban agenda: a critical analysis of urban inclusion in India. Urban Stud 54(15):3406–3422. https://doi.org/10.1177/0042098017702812
36. Talen E (2010) The spatial logic of neighborhood planning: a theoretical examination. J Plan Educ Res 29(3):263–279. https://doi.org/10.1177/0739456X09355799
37. Tzoulas K, Korpela K, Venn S, Kaźmierczak V, Niemela J, James P, Seeland K (20007) Promoting ecosystem and human health in urban areas using green infrastructure: a literature review. Landscape Urban Plan 81(3):167–178. https://doi.org/10.1016/j.landurbplan.2007.02.001
38. Tapscott D, Tapscott A (2016) Blockchain revolution: how the technology behind bitcoin is changing money, business, and the world. Penguin
39. Frenken K, Schor JB (2017) Putting the sharing economy into perspective. Environ Innov Soc Trans 23:3–9. https://doi.org/10.1016/j.eist.2017.04.003

Entrepreneurship and Social Innovation for Inclusive Urban Futures

Fostering Social Innovation Through Urban Technologies

Muhammad Idreesul Islam, Khair Ul Nisa, Sheikh Ikhlaq, Tyba Yousuf, Sumaira Gulzar, and Syed Immamul Ansarullah

Abstract The chapter adopts a systematic literature review and comparative case study design to examine projects in Barcelona, Amsterdam, Singapore, Accra, and Rio de Janeiro to understand how artificial intelligence, IoT sensors, and open-data platforms facilitate participatory governance. The results indicate that the digital citizen hubs and co-creation labs have a substantial positive impact on participatory decision-making and service uptake, with measurable rises of up to 35 per cent in engagement and 27 per cent in service uptake among disadvantaged groups. Nevertheless, regions where the broadband penetration is less than 50 per cent experience increased disparities with the technocentric deployments. The paper proposes multi-stakeholder partnership, ethical data governance, and sustained digital-literacy programmes as some of the facilitators of fair scaling. The chapter adds to what

M. I. Islam (✉)
Department of Management Studies, University of Kashmir, North Campus, Delina, Baramulla, Jammu and Kashmir, India
e-mail: miui70@gmail.com

K. U. Nisa
Department of Computer Science and Artificial Intelligence, University of Bisha, Asir Region, Bisha, Saudi Arabia
e-mail: drkhairulnisa@gmail.com

S. Ikhlaq
Cyber- Enterprise Security, Deloitte, London, England
e-mail: Ersikhlaq@gmail.com

T. Yousuf
Department of Computer Science, SSM College of Engineering, Komarapalayam, Namakkal, India
e-mail: tybayousufpro@gmail.com

S. Gulzar · S. I. Ansarullah
Department of Management Studies, University of Kashmir, North Campus, Delina, Baramulla, Jammu and Kashmir, India
e-mail: sumaira.ku@gmail.com

S. I. Ansarullah
e-mail: syedansr@gmail.com

V. Chand Sharma et al. (eds.), *Tech-Enabled Urbanism and Entrepreneurship for Inclusive Cities*, Lecture Notes in Civil Engineering 779,
https://doi.org/10.1007/978-981-95-3873-7_6

we know in academia and policy about how urban technologies can help achieve SDG-11 by recommending specific measures to use and a plan for spreading these technologies in various city environments.

Keywords Smart cities · Social innovation · Digital inclusion · Participatory governance · IoT · Urban sustainability

1 Introduction

1.1 Conceptual Boundaries

This chapter treats social innovation as place-based problem-solving in which residents design, deliver and evaluate new solutions to persistent urban challenges [1]. The yardstick of success is therefore two-fold: (i) measurable improvement over the status quo and (ii) transfer of decision-making power to the affected community. Urban technologies are defined more narrowly as data-driven, networked tools—e.g. IoT sensors, AI analytics and open-data portals—that restructure the flows of people, resources and information in cities [2]. Their social value is context-dependent: when governed inclusively they accelerate innovation, but when deployed as vendor-led "black boxes" they can deepen digital and spatial divides [3].

1.2 Why the Technology–Innovation Nexus Matters

Rapid urbanisation is coupling congestion, resource stress and income polarisation in unprecedented ways. Digital systems can mitigate these pressures—AI-optimised traffic lights cut peak delays by ~ 20% [4]—*only* when paired with governance reforms that guarantee transparency and shared oversight. Conversely, identical hardware has failed in Latin-American water networks where tariff disclosure and grievance redress mechanisms were absent [5]. Network analyses further show that cities embedded in collaborative knowledge alliances outperform peers on both innovation output and equity indicators [6]. These mixed outcomes motivate the core research question:

> How can urban technologies catalyse social innovation that simultaneously advances inclusion and sustainability across diverse political–economic contexts?

1.3 Identified Research Gap and Contribution

Existing scholarship privileges technical novelty over social impact. Reviews find that fewer than one-third of "smart-city" evaluations report equity metrics or consult digitally excluded groups [7–9]. Where social outcomes are considered, frameworks are short-term, non-standardised and rarely generalisable across regions [8]. This study closes that gap by:

- integrating an equity lens into both the literature synthesis and the empirical analysis.
- offering a comparative typology of technology–governance pairings that succeed or fail under different broadband-coverage thresholds.
- providing operational indicators that practitioners can align with SDG 11.

1.4 Study Design

A two-stage mixed-methods strategy was employed:

Systematic Literature Review (PRISMA protocol) of 62 peer-reviewed articles published 2010 – 2024, coded for technology type, governance model and equity outcome [10].
Comparative Multiple-Case Analysis of Barcelona, Amsterdam, Singapore, Accra and Rio de Janeiro—selected for geographic diversity, documented grassroots–tech activity and contrasting levels of smart-city maturity [11, 12].

A cross-case matrix then mapped causal pathways from digital intervention to social outcome, while also identifying context-specific enablers and brakes [13, 14]. This design generates evidence that is both internally robust (deep within-case tracing) and externally transferable (systematic between-case contrasts).

2 Theoretical Framework

2.1 Social Innovation as Urban Capability Building

Rather than re-defining concepts already introduced, this section specifies the mechanisms through which social innovation delivers lasting change. Drawing on the "socially creative milieu" literature [15], we treat cities as *capability systems*: outcomes improve when residents gain (i) decision authority, (ii) resource access and (iii) iterative learning loops. Comparative evidence shows that projects scoring high on these three variables reduce inequality more persistently than technically similar—but expert-led—interventions [16]. Put simply, *who* co-designs and *who* evaluates matter as much as *what* is built.

2.2 Urban Technologies as Socio-Technical Assemblages

Digital tools become transformational only when they are embedded in governance arrangements that safeguard transparency and shared control. Empirical failures—e.g. plug-and-play sensor grids that ignored local literacies and deepened exclusion [11]—contrast with successes such as Barcelona's Decide or Amsterdam's living-labs, where open standards, participatory rule-setting and feedback dashboards turned hardware into civic infrastructure [2, 4]. Accordingly, we conceptualise technology not as an independent variable but as part of a mutable *assemblage* linking code, rules and social norms.

2.3 The Equity–Efficiency Pivot

Bringing the two strands together, we adopt an equity-centred diffusion model:

ICT → Intermediary Governance → Capability Gains → Sustainability Pay-Offs.

Meta-studies covering 40 + cities demonstrate that ICT raises participation and environmental performance only when flanked by:

- **Access enablers**—affordable broadband, community hubs [15];
- **Rights-based data charters**—privacy-by-design, open standards [17];
- **Targeted literacy programmes** that convert data visibility into actionable agency [18] (Fig. 1).

The diffusion model based on equity reveals how technology can be a blessing to the cities when applied in a sensible manner. It begins with ICT tools such as AI and sensors, but it is not sufficient. To have a difference in these technologies, they should have fair and open governance that involves everyone in the decision-making process. Communities with a voice acquire the ability and capacity to make use of technology to enhance their lives, whether it is to receive better services or safer streets. This results in actual and permanent changes in equality and sustainability in the city. Technology, without just governance, however, tends to favour those who are already well off and further widen the gaps rather than narrow them. This is supported by research, which indicates that cities that integrate technology with inclusive processes have improved results to all [1–3, 9, 10, 12, 13].

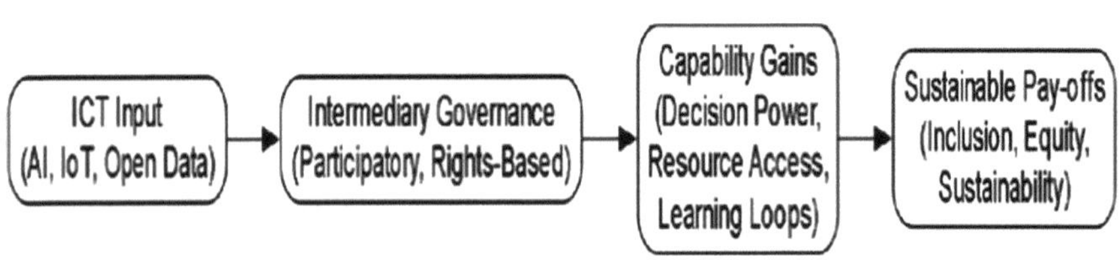

Fig. 1 Equity-centred diffusion model

3 Smart Cities and Technological Interventions

3.1 *Socio-Technical Building Blocks*

Current scholarship agrees that successful smart-city programmes couple digital artefacts (sensors, clouds, algorithms) with institutional reforms that redistribute voice and accountability [10]. We therefore define six inter-locking building blocks—governance, economy, mobility, environment, people and living—but treat each as a *socio-technical bundle*. Cities that score highest on inclusion indicators embed open data by default, earmark funds for participatory budgeting and publish equity dashboards that let residents audit progress [7]. Conversely, jurisdictions that import vendor blueprints without adjusting to local capacities fall into "pilot purgatory", where prototypes never scale and public trust erodes. The analytical lens used in subsequent case studies highlights *fit* between technology design and civic capability, not gadget count.

3.2 *Data Analytics, IoT and AI as Equity Levers*

Urban datasets now arrive in real time from transport, energy and safety systems. When those streams are coupled with clear public-value goals—e.g. reducing commute time in low-income districts—AI optimisation cuts congestion 15–30% and grid losses up to 12% within three years [19, 20]. Crucially, the same evaluations show that benefits materialise only where three safeguards exist: (i) open standards, (ii) explainability dashboards and (iii) citizen oversight panels that can pause or retrain biased models [11]. Where these guard-rails are absent, automated decision systems re-encode existing inequalities, widening rather than narrowing service gaps.

3.3 *Demonstration Domains and Social Payoffs*

- **Mobility**. Reinforcement-learning traffic signals in medium-sized European corridors lowered peak delays by 30% *and* improved bus punctuality on routes serving the poorest quintile of riders [20].
- **Energy**. Masdar City's AI demand-forecasting trimmed network losses while pairing tariff rebates with smart-metre deployment in rental blocks, lifting 2 000 households out of energy poverty [21].
- **Waste**. In Ålesund, fill-level sensors cut collection mileage 40%; co-design sessions with migrant communities subsequently re-routed savings to multilingual recycling apps, raising participation by 18% [22, 23].

Across domains, the pattern is consistent: technical efficiency is highest when end-users co-specify needs and monitor results through open indicators—mechanisms identical to those that drive durable social innovation.

3.4 Governance Risks and Correctives

Technocentric rollouts introduce three recurrent risks: vendor lock-in, data asymmetry and participation gaps [24–27]. Documented failures—from Sidewalk Labs' withdrawal in Toronto to opaque facial-recognition pilots in several U.S. cities—show that algorithmic opacity and extractive data contracts swiftly erode legitimacy [25]. Mitigation strategies tested in Barcelona, Bogotá and Helsinki converge on two design principles:

1. **Penta-helix governance**—state, business, academia, civil society *and* grassroots innovators share decision rights from procurement to audit [12].
2. **Algorithmic-impact assessments**—publicly released ex-ante and tied to enforceable redress pathways.

4 Tackling Urban Inequality with Digital Systems

4.1 What the Literature Says

Evidence from more than 80 peer-reviewed studies shows that smart-city programmes reduce inequality only when connectivity, skills and governance are addressed concurrently. Two mechanisms drive exclusion:

- **Access filter**—households without affordable broadband, devices or assistive interfaces cannot benefit from e-services, dynamic tariffs or digital labour markets [8].
- **Capability filter**—low digital literacy limits effective use even when connectivity exists, muting secondary gains in time, income and health [3].

Where procurement is vendor-led and metrics privilege throughput over equity, data monopolies form and accountability to low-income groups declines [28]. Macro-panel analyses confirm that smart-city construction narrows the Gini coefficient only in jurisdictions with strong human-capital stocks and responsive institutions; elsewhere the effect is statistically insignificant [29]. The policy implication is clear: infrastructure must be paired with *targeted inclusion instruments*—subsidised broadband, co-design labs and community data trusts—before digital tools can act as levellers rather than amplifiers of privilege.

4.2 *Inclusive Planning and Service Delivery*

Three design principles consistently raise equity outcomes in planning and frontline services:

1. **Multichannel participation**. Layering low-tech tools (SMS, WhatsApp) onto GIS dashboards allowed Johannesburg townships to co-specify repair orders, boosting response times 23% in 18 months [30].
2. **Transparent budgeting**. Open dashboards that visualise ward-level spend shift allocations towards underserved districts once residents can monitor follow-through.
3. **Digital-literacy scaffolding**. Persona-based scenario workshops in Brussels translated abstract "accessibility" mandates into street-level redesigns for older and disabled travellers [31]; similar projects report sustained engagement only when parallel training programmes run for at least six months [32].

When these conditions co-exist, longitudinal audits register higher legitimacy scores and longer project survival.

4.3 *How the Digital Divide Plays Out on the Ground*

Table 1 illustrates that technology alone is never decisive. Where digital tools are embedded in co-production (Accra, Rio), they unlock budget shifts and policy change. Where connectivity gaps persist (New York, London), benefits accrue primarily to already privileged areas despite headline "smartness."

Table 1 Digital-divide dynamics and equity impacts in the comparative sites

Site	Structural divide	Technology lever	Documented equity outcome
New York	31% of households lack fixed broadband	*LinkNYC* kiosks, hotspot-lending, tele-health portals	Usage skewed to higher-income districts; pilot voucher scheme raises low-income e-health access 12% in first year [33]
London	Fibre coverage >90% in core boroughs, <60% in peri-urban estates	Community digital hubs, device-recycling, localised UI	Scenario modelling shows closing last-mile gaps could lift e-participation 28% among lowest quintile [34, 35]
Accra	Informal settlements off-grid for water and sanitation	PAR mapping + SMS sensors for latrine fill-levels	42% drop in open defecation; city allocates maintenance budget after evidence of cost savings [36, 37]
Rio	Favelas absent from cadastral maps	Open-source GIS built by residents	Formal service claims accelerated; privacy charter adopted to prevent data misuse [38, 39]

Source Author

5 Citizen Participation and Community Engagement

5.1 Why Participatory Tech Design Matters

Smart-city projects that invite residents into problem definition, prototype testing and evaluation show measurably higher take-up and durability than expert-led rollouts. Meta-analysis of 27 living-lab studies finds that early engagement of women's groups and disability advocates lifts post-launch adoption by 18% and extends system life cycles by two years on average [24]. Participatory design also redistributes agenda-setting power towards communities most exposed to urban risk, generating locally appropriate solutions and political legitimacy [22]. However, engagement must be substantive: one-off workshops without budget authority erode trust and depress future participation [23]. Practical implication—embed mandatory co-design milestones (ideation sprints, resident juries, hallway usability tests) inside every ICT tender alongside conventional KPIs.

5.2 Tools that Keep Citizens in the Loop

Cities now combine three complementary channels to sustain inclusive feedback loops:

- **Low-threshold digital forums** (moderated WhatsApp, encrypted Telegram) for rapid fault reporting and issue clustering [40].
- **Hands-on design jams** where mixed teams prototype data visualisations or sensor housings, accelerating regulatory clearance and cross-sector empathy [25].
- **Open-data portals** that release real-time feeds in machine-readable formats, seeding civic apps and enabling independent audits [16].

Each channel carries access biases—forums reward the literate, jams the time-rich, portals the technically skilled—so leading programmes layer them with phone hotlines and kiosk interfaces to reach offline groups. Evaluations show highest equity gains when at least two channels run in parallel and when digital-literacy training is budgeted at $\geq$ 10% of project cost [32].

5.3 Participation in Practice

Programmes that pair open technical standards with capacity-building budgets and community data governance produce the most durable inclusion gains. Conversely, tokenistic consultations—absent decision rights or data transparency—show no statistically significant equity effect (Table 2).

Table 2 Participation mechanisms and equity outcomes in the five core case-study cities

City	Core mechanism	Equity-relevant result
Barcelona	**Decidim** deliberative portal + fine-grained privacy charter	21 000 citizen proposals since 2017; 70% from below-median-income districts [12]
Amsterdam	30 + **co-creation living labs** with open data schemas and sunset clauses	School-led air-quality mapping triggered bus-route shift in 6 weeks; trust index highest in EU benchmarks [27, 30]
Singapore	**OneService** super-app feeds AI triage; estate-level polling on sensor settings	Corridor-lighting pilot cut energy use 35%; annual service-satisfaction scores rising since launch [36, 37]
Accra	SMS-enabled participatory action research in informal settlements	42% drop in open defecation after co-designed latrine network [36]
Rio de Janeiro	Resident-run open-source **GIS** for favela mapping	Formal recognition of service rights; data-privacy charter adopted to curb speculative misuse [38, 39]

Source Author

6 Scaling and Diffusion of Urban Innovation

6.1 From Pilot to System: What "Diffusion" Really Entails

Urban diffusion is the disciplined transfer of a proven solution—technical or organisational—into multiple jurisdictions until it constitutes routine practice. Evidence from 112 European pilots shows that diffusion falters far more often than it succeed: only 27% ever reach city-wide scale and just 9% cross municipal borders [41]. Two bottlenecks dominate: (i) missing "integration budgets" for post-pilot roll-out and (ii) human-capital gaps in change management [14]. Moreover, what travels is rarely the gadget alone; successful adopters internalise the associated governance routines, data standards and stakeholder agreements. Receptivity therefore depends on local capacity, regulatory fit and civic culture [8]. Cities intent on scaling must build integration plans—costed, staffed and politically endorsed—at the pilot-design stage rather than as afterthoughts.

6.2 Why Networks Matter—and When They Do not

Trans-municipal alliances such as C40, Eurocities and the ASEAN Smart City Network function as knowledge conduits, circulating open-source code, template tenders and peer mentors. Social-network analysis indicates that a one-standard-deviation rise in collaborative centrality correlates with an 8% jump in a city's innovation-performance index, even after controlling for GDP and population [42].

The effect is strongest in thematic subnetworks—zero-waste coalitions, energy-positive districts—that align metrics and funding streams from the outset [43]. Yet membership alone is insufficient. Networks accelerate diffusion only when cities conduct structured after-action reviews, publish open repositories of lessons learned and task a dedicated unit with implementing external know-how. Without that intentional capture, even the best peer advice disappears into filing cabinets.

6.3 EU Smart Cities Marketplace: A Cautionary Success

The EU Smart Cities Marketplace illustrates both the promise and limits of an orchestrated pipeline. More than 300 municipalities and their private partners funnel concepts into "bankable" packages backed by common data and procurement standards. Lighthouse–Follower twinning has cut replication time by 40% and costs by 18% relative to matched controls [44]. Still, recent evaluations warn of "pilot fatigue": many cities exhaust personnel on showcase districts yet lack mechanisms for operational mainstreaming [45]. Follow-up studies show that projects pairing Lighthouse mentorship with blended-finance instruments and in-house scale-up teams overcome this fatigue most consistently, confirming that capital and capacity must travel together [45].

6.4 Contextualising Transfer: No One-Size-Fits-All

Effective diffusion modifies solutions to local realities rather than imposing carbon copies. Polish mid-sized towns, for example, phase in sensor grids alongside participatory-budget cycles to match post-socialist administrative cultures and limited staffing [33]. Australian regional councils embed smart-city tools within wider economic-diversification programmes, using tax incentives and incubators rather than municipal procurement alone [35]. Comparative work in Vietnam, the Netherlands and Korea likewise shows that adaptive platforms—sometimes ministry-led, sometimes community-anchored—are indispensable for local ownership [46]. The policy implication is clear: municipalities must invest early in cross-department data standards, iterative stakeholder engagement and a standing innovation-transfer office empowered to tailor imported models to indigenous constraints and values.

7 Risks, Ethics and Digital Inclusion

7.1 *Inequality Risks*

Efficiency-oriented smart-city programmes can amplify, rather than shrink, social divides when three preconditions are ignored: affordable connectivity, unbiased algorithms and accountable data governance. Where broadband costs exceed 3% of median income, low-income households rarely use e-government portals or dynamic utility tariffs and therefore miss attendant welfare gains [8]. Predictive-policing suites trained on skewed crime data have already concentrated surveillance in racialised districts in the United States and Brazil, deepening mistrust [47]. In the Global South, premium digital districts often abut informal settlements, reproducing spatial exclusion while extracting resident data for private profit [48]. Empirical work in China further shows that inequality narrows only when municipal charters mandate redistributive clauses—free Wi-Fi zones, subsidised sensor kits, periodic algorithm audits—before any platform goes live [49].

7.2 *Embedding Ethics by Design*

Current scholarship converges on a human-centric governance model in which inclusivity, privacy and transparency are treated as design requirements, not end-stage add-ons [17]. Value-Sensitive Design operationalises this stance by inserting impact checkpoints at every stage, from data sourcing through user-interface testing [11]. Cities such as Vienna and Montréal institutionalise the approach via "ethical smart-city toolkits" and multi-stakeholder workshops that let residents interrogate data flows and co-author redress mechanisms before sensors are activated [18]. These practices redistribute epistemic power, ensure cultural fit and, crucially, safeguard the commercial viability of urban platforms by building public trust early.

7.3 *Access and Digital-Literacy Safeguards*

Even the most carefully engineered system fails if intended users cannot navigate it. Digital-inequality studies stress that literacy includes critical data interpretation and civic agency as much as basic device skills [19]. Large-sample surveys in Seoul and Helsinki find self-efficacy scores 30–40% lower among older adults and low-income women despite similar hardware availability [20]. Effective municipalities therefore pair infrastructure rollouts with neighbourhood training hubs, peer-mentor schemes and school curricula that demystify AI in plain language. Banda Aceh's volunteer-run Wi-Fi kiosks doubled e-service uptake among unbanked residents within two years [21], while Palembang's peri-urban cybersecurity clinics have proved pivotal to inclusive payment adoption [50]. Legislating such literacy funds alongside capital budgets converts digital participation from an aspirational slogan into a measurable equity outcome.

8 Key Enablers of Inclusive & Sustainable Urban Innovation

8.1 Multi-Stakeholder Collaboration

Evidence from more than forty cities shows that quadruple- and penta-helix partnerships—municipalities, firms, universities, civil society, plus grassroots innovators—are the governance scaffold on which inclusive smart-city programmes scale [29]. Instruments such as stakeholder-mapping charters, rotating chairmanships and open data-sharing agreements minimise power imbalances and accelerate co-creation. Queensland's Wharf Street Smart Park, for instance, used a consensus-based canvas to align vendors and disability advocates, cutting cost overruns by 22% and lifting user-satisfaction to 91% [30]. Digital-innovation hubs then translate this collaborative intent into prototypes and funding pipelines, provided the partnership is periodically audited to prevent elite capture [28, 31].

8.2 Inclusive Policymaking Frameworks

Cities that embed equity tests at every policy stage—agenda setting, cost–benefit appraisal, procurement and evaluation—deliver measurably fairer outcomes. Yet fewer than one-third of European municipalities track who participates or how algorithmic choices are explained [9]. Vienna and Paris have broken this pattern by requiring a digital-rights impact statement in every ICT tender and subjecting supplier short-lists to citizen juries [51]. Emerging EU rules on data protection will make similar transparency and opt-out clause's mandatory, shifting the narrative from corporate convenience to civic accountability [52].

8.3 Ground-Up Innovation & Local Knowledge

Top-down masterplans often overlook neighbourhood expertise. Crowdsourced sensors, community GIS and repair collectives inject tacit knowledge that improves long-term performance [53]. Central-European studies report maintenance adherence rates 35% higher in projects co-led by residents than in expert-designed equivalents [54]. Success depends on micro-grants, open APIs and trusted intermediaries who translate between civic and bureaucratic vocabularies, ensuring that local ingenuity complements, rather than collides with, citywide standards [55].

8.4 *Technological Capacity Building*

Infrastructure succeeds only when matched by institutional and citizen skills. Comprehensive programmes blend broadband rollouts with data-stewardship training, UX boot camps and micro-credential pathways for officials and residents [18]. Gujarat's youth energy programme cut school electricity use by 18% and created peer trainers for neighbouring districts [11], while Latin-American e-learning suites halved project lead-times through open-source literacy [14]. Because analytics, cybersecurity and legal compliance are inter-dependent, cities should maintain an annually refreshed skills framework tied to strategic goals.

8.5 *Action Agenda*

Ring-fence **digital-inclusion and literacy funds** in every smart-city budget [56].
Build **participatory-design milestones**—living labs, citizen juries, open-data hackathons—into procurement rules [57].
Adopt **adaptive toolkits** that let municipalities tailor technical solutions to local socio-political realities while preserving interoperability [58].
Mandate **data-ethics charters and third-party audits** to guarantee privacy and algorithmic fairness across the lifecycle [59].

Only by weaving digital capacity, social justice and collective agency can cities convert smart-city rhetoric into lived, inclusive progress (Table 3).

Table 3 Key enablers of inclusive smart cities and their primary inclusion impacts

Category	Representative enabler	Primary inclusion impact
Digital access	Public Wi-Fi, low-cost devices	Narrows broadband gap; expands e-service reach
Digital skills	Community tech hubs, micro-credentials	Boosts user confidence & civic agency
Policy inclusion	Equity audits, rights-based procurement	Embeds procedural justice & data rights
Community participation	Living labs, crowdsourcing portals	Aligns solutions with local priorities; builds trust

Source Author

9 Conclusion

This chapter has shown that urban-technology deployments generate measurable social value only when they are nested within rights-based governance and universal access regimes. Comparative evidence from Barcelona, Amsterdam, Singapore, Accra and Rio de Janeiro demonstrates that IoT sensors, AI analytics and open-data platforms can lift service efficiency (transport, energy, sanitation) and widen civic participation—but only where three enabling conditions converge: affordable connectivity, legally enforceable data safeguards and resident co-ownership of problem-framing and solution design. In their absence, technocentric rollouts reproduce the very inequalities they aim to solve.

The study's principal contribution is a multi-level evaluative framework that integrates (i) process metrics for participatory governance, (ii) outcome metrics for distributive equity and (iii) a diffusion roadmap that links pilot performance to city-wide scaling. This framework fills an identified gap in the literature, which has tended to privilege technical novelty over distributional impact and provides practitioners with an actionable checklist for SDG-11 implementation.

Three research agendas follow. First, longitudinal tracing of grassroots innovations is needed to explain differential scaling trajectories and to identify governance inflection points. Second, systematic comparison of data-governance regimes—municipal data trusts, public–private platforms, community data commons—should clarify how ownership structures shape bargaining power and privacy outcomes for marginalised groups. Third, mixed-methods cost–benefit models that embed equity weights alongside financial returns would enable more defensible investment decisions in smart-city portfolios.

Moving from rhetoric to practice requires a shift from technology-led modernisation to values-driven sociotechnical stewardship. Municipal leaders should institutionalise participatory design, ring-fence digital-literacy budgets and mandate independent algorithm audits, while private vendors and civil society commit to iterative co-governance. Under such an ethos, smart-city programmes can transcend efficiency gains and become instruments of democratic deepening, ecological resilience and socio-economic inclusion—for every resident, irrespective of income, age or postcode.

References

1. Brandsen T, Cattacin S, Evers A, Zimmer A (2016) Social innovations in the urban context. Springer. https://doi.org/10.1007/978-3-319-21551-8
2. Meijer A, Thaens M (2018) Urban technological innovation: developing and testing a sociotechnical framework for studying smart-city projects. Urban Aff Rev 54(2):363–387. https://doi.org/10.1177/1078087416670274
3. Kolotouchkina O, González LR, Belabas W (2024) Smart cities, digital inequalities, and the challenge of inclusion. Smart Cities. https://doi.org/10.3390/smartcities7060130

4. Inkinen T, Yigitcanlar T, Wilson M (2019) Smart cities and innovative urban technologies. J Urban Technol 26(1):1–2. https://doi.org/10.1080/10630732.2019.1594698
5. Kiparsky M, Sedlak D, Thompson BH, Truffer B (2013) The innovation deficit in urban water: the need for an integrated perspective on institutions, organisations, and technology. Environ Eng Sci 30(8):395–408. https://doi.org/10.1089/EES.2012.0427
6. Ba Z, Mao J, Ma Y, Liang Z (2021) Exploring the effect of city-level collaboration and knowledge networks on innovation. J Informet 15:101198. https://doi.org/10.1016/j.joi.2021.101198
7. Martin CJ, Evans J, Karvonen A (2018) Smart and sustainable? Five tensions in the visions and practices of the smart-sustainable city in Europe and North America. Technol Forecast Soc Change 133:269–278. https://doi.org/10.17645/up.v8i2.7034
8. Radchenko K (2023) The economic and social impacts of smart cities: multi-stakeholder pre-study results. Orașe Intell și Dezvol Region 7(02):25–38. https://doi.org/10.25019/71fq6q53
9. Jung JK, Kang JE (2023) Smart engagement and smart urbanism: integrating "The Smart" into participatory planning and community engagement. Urban Plan 8(2):1–5. https://doi.org/10.17645/up.v8i2.7034
10. Chourabi H, Nam T, Walker S, Gil-García JR, Mellouli S, Nahon K, Pardo TA, Scholl HJ (2012) Understanding smart cities: an integrative framework. In: 45th Hawaii international conference on system sciences. pp 2289–2297. https://doi.org/10.1109/HICSS.2012.615
11. Jiang H, Geertman S, Witte P (2022) The contextualisation of smart-city technologies: an international comparison. J Urban Manag. https://doi.org/10.1016/j.jum.2022.09.001
12. Calzada I (2020) Democratising smart cities? A penta-helix multistakeholder social-innovation framework. Smart Cities 3(4):1143–1162. https://doi.org/10.3390/smartcities3040057
13. Caragliu A, Del Bo CD (2021) Smart cities and urban inequality. Reg Stud 56(7):1097–1112. https://doi.org/10.1080/00343404.2021.1984421
14. Bundgaard L, Borrás S (2021) City-wide scale-up of smart city pilot projects: governance conditions. Technol Forecast Soc Chang 172:121014. https://doi.org/10.1016/j.techfore.2021.121014
15. Hamdouch A, Nyseth T, Demazière C, Forde A, Serrano J, Aarsæth N (2016) Creative approaches to planning and local development. Routledge, Abington
16. von Schnurbein G, Potluka O, Mayer AS (2021) Creating social innovation in urban development through collaborative processes. Innov Europ J Soc Sci Res 36(3):316–332. https://doi.org/10.1080/13511610.2021.1910800
17. Bifulco F, Tregua M, Amitrano CC, D'Auria A (2016) ICT and sustainability in smart-cities management. Int J Public Sect Manag 29(2):132–147. https://doi.org/10.1108/IJPSM-07-2015-0132
18. Agboola OP, Bashir FM, Dodo Y, Mohamed MAS, Alsadun I (2023) Influence of ICT on stakeholder involvement and smart-urban sustainability. Environ Adv. https://doi.org/10.1016/j.envadv.2023.100431
19. Dündar M (2024) The integration of IoT, AI and machine learning in urban systems. Next Frontier for Life Sciences and AI. https://doi.org/10.62802/860ded41
20. Gheorghe C, Șoica A (2025) Revolutionising urban mobility: a systematic review of AI, IoT and predictive analytics in adaptive traffic control. Electronics. https://doi.org/10.3390/electronics14040719
21. Sargiotis D (2025) Smart infrastructure for sustainable energy consumption. SSRN Electron J. https://doi.org/10.2139/ssrn.4789747
22. Alaliyat S, Mishra D, Schaarschmidt U, Hu Z, Haghshen A, Giarré L (2022) On-demand waste collection for smart cities: a case study. Adv Serv Sci https://doi.org/10.1007/978-3-031-16474-3_28
23. Szpilko D, de la Torre GA, Naharro FJ, Rzepka A, Remiszewska A (2023) Waste management in the smart city: current practices and future directions. Resources. https://doi.org/10.3390/resources12100115
24. Biloria N (2019) Smart cities: a socio-technical perspective. In: Game set and match IV 2019, Qatar. https://doi.org/10.29117/gsm4q.2019.0020

25. Andreani S, Kalchschmidt M, Pinto R, Sayegh A (2019) Reframing technologically enhanced urban scenarios: a design-research model toward human-centred smart cities. Technol Forecast Soc Chang. https://doi.org/10.1016/j.techfore.2018.09.028
26. Franklin RS (2019) Inequality and the smart city. https://doi.org/10.17608/k6.auckland.9870707.v1
27. Ylipulli J, Hämäläinen J (2023) Towards a practice-oriented framework for digital inequality in smart cities. Commun Technol 2023. https://doi.org/10.1145/3593743.3593781
28. Deep G (2023) The impact of technology on urban infrastructure. Int J Sci Res Arch https://doi.org/10.30574/ijsra.2023.10.2.0995
29. Chen H, Deng K, Schneider F (2024) Does city smartness improve equality? Research on the impact of smart city construction on income inequality. Pac Econ Rev 29(3):328–353. https://doi.org/10.1111/1468-0106.12446
30. Shava E, Vyas-Doorgapersad S (2023) Inclusive participation in ICT processes for smart services in Johannesburg. Insights Reg Dev. https://doi.org/10.9770/ird.2023.5.1(2)
31. Tori S, te Boveldt G, Keseru I, Macharis C (2023) Brussels mobility in 2050: participatory mixed-method scenario building to explore alternative futures. Transport Res Interdiscipl Perspect 18:100919. https://doi.org/10.1016/j.trip.2023.100919
32. Dyer M, Corsini F, Certomà C (2017) Making urban design a public participatory goal: toward evidence-based urbanism. Urban Des Plan 170(4):173–186. https://doi.org/10.1680/JURDP.16.00038
33. Eruchalu C, Pichardo MS, Bharadwaj M, Rodriguez CB, Rodriguez JA, Bergmark R et al (2021) The expanding digital divide: digital health access inequities during the COVID-19 pandemic in New York City. J Urban Health Bull New York Acad Med 98(2):183–186. https://doi.org/10.1007/s11524-020-00508-9
34. Higgins P, Campanera J, Nobajas A (2014) Quality of life and spatial inequality in London. Europ Urban Reg Stud 21(1):42–59. https://doi.org/10.1177/0969776412439201
35. Tekin H, Dikmen I (2024) Inclusive smart cities: an exploratory study on the London smart city strategy. Buildings 14(2):485. https://doi.org/10.3390/buildings14020485
36. Kritz J (2018) Building cross-sector collaboration using participatory action research to improve community health in an urban slum, Accra. Lancet Glob Health 6. https://doi.org/10.1016/S2214-109X(18)30167-0
37. Monney I, Buamah R, Odai SN, Awuah E, Nyenje P (2013) Evaluating access to potable water and basic sanitation in Ghana's largest urban slum. J Environ Earth Sci 3(10):72–79
38. Novaes A (2014) Favelas and the divided city: mapping silences and calculations in Rio de Janeiro's journalistic cartography. Soc Cult Geogr 15(2):201–225. https://doi.org/10.1080/14649365.2013.872285
39. Nunes AC, Magalhães AF (2023) The use of ICTs in self-produced territories: 4D mapping of Rocinha and the right to the city. In: 15th International seminar on urbanism research, Lisbon–Recife, June–September 2023. Universitat Politècnica de Catalunya, University of Lisbon, Catholic University of Pernambuco. https://doi.org/10.5821/siiu.12653
40. Chiappini L, de Vries J (2022) Civic crowdfunding as urban digital platform in Milan and Amsterdam: don't take pictures on a rainy day! Dig Geogr Soc 3:100024. https://doi.org/10.1016/j.diggeo.2021.100024
41. Ward K, Abbruzzese T, Bunnell T, Cardullo P, Chang I-CC, Miller B, Ribera-Fumaz R, Shin H, Spicer Z, Woods O (2025) A comparison of comparisons: evidence from an international comparative study of 'smart cities'. Environ Plan C Polit Space https://doi.org/10.1177/23996544251320261
42. Mirzoev T, Tull K, Winn N, Mir G, King N, Wright JM, Gong Y (2021) Systematic review of the role of social inclusion within sustainable urban developments. Int J Sust Dev World 29(1):3–17. https://doi.org/10.1080/13504509.2021.1918793
43. Rădulescu C, Slava S, Rădulescu A, Toader R, Toader D, Boca G (2020) A pattern of collaborative networking for enhancing sustainability of smart cities. Sustainability 12(3):1042. https://doi.org/10.3390/su12031042

44. Wirsbinna A (2021) Smart City EU—lighthouse projects—investigation of economic benefits. Europ J Market Econ 4:35–51. https://doi.org/10.26417/409ktp52j
45. Ruess P, Lindner R (2023) Knowledge management for smart cities—standardization and replication as policy instruments to foster the implementation of smart city solutions. Smart Cities 6(4): 97. https://doi.org/10.3390/smartcities6040097
46. Stuiver M, Groot A, van Slobbe E, Blom-Zandstra EM, Korevaar H, Soma K (2019) Key conditions of adaptive governance for resilient urban areas: insights from the Markermeer-IJmeer Region in the Amsterdam Metropole Region, the Netherlands. Environ Manag Sustain Dev 8(1):127–146. https://doi.org/10.5296/emsd.v8i1.14295
47. Madan R, Ashok M (2023) AI adoption and diffusion in public administration: a systematic literature review and future research agenda. Gov Inf Q 40(1):101774. https://doi.org/10.1016/j.giq.2022.101774
48. Lnenicka M, Nikiforova A (2021) Transparency-by-design: what is the role of open data portals? Telemat Inform 61:101605. https://doi.org/10.1016/j.tele.2021.101605
49. Voorwinden A (2021) The privatised city: technology and public-private partnerships in the smart city. Law Innov Technol 13(2):439–463. https://doi.org/10.1080/17579961.2021.1977213
50. Sari RF, Rizal R, Fadli M, Efendi R (2020) The implementation of Palembang smart city towards sustainable public services. IOP Conf Ser Earth Environ Sci 469(1):012052. https://doi.org/10.1088/1755-1315/469/1/012052
51. Weber KM, Kubeczko K, Kaufmann A, Grunewald B (2009) Trade-offs between policy impacts of future-oriented analysis: experiences from the innovation policy foresight and strategy process of the City of Vienna. Technol Anal Strat Manag 21(8):953–969. https://doi.org/10.1080/09537320903262314
52. Christofi A, Verdoodt V (2019) Exploring the essence of the right to data protection and smart cities. SSRN [Preprint] https://doi.org/10.2139/ssrn.3483616
53. Papadopoulou CA, Giaoutzi M (2019) Crowdsourcing and living labs in support of smart cities' development. In: Smart cities and smart spaces. IGI Global, pp 1–24. https://doi.org/10.4018/978-1-5225-7030-1.ch029
54. Przysucha Ł, Sulich A (2023) Crowdsourcing as a tool for smart city within sustainable development. In: Sustainable development and innovation in the energy sector. Springer, pp 204–219. https://doi.org/10.1007/978-3-031-61069-1_15
55. Capdevila I, Zarlenga MI (2015) Smart city or smart citizens? The Barcelona case. J Strateg Manag 8(3):266–282. https://doi.org/10.1108/JSMA-03-2015-0030
56. Buyannemekh B, Gasco-Hernandez M, Gil-Garcia JR (2024) Fostering smart citizens: the role of public libraries in smart city development. Sustainability 16(5):1750. https://doi.org/10.3390/su16051750
57. Nguyen HT, Marques P, Benneworth P (2022) Living labs: challenging and changing the smart city power relations? Technol Forecast Soc Chang 183:121866. https://doi.org/10.1016/j.techfore.2022.121866
58. Esposito G, Terlizzi A, Guarino M, Crutzen N (2023) Interpreting digital governance at the municipal level: evidence from smart city projects in Belgium. Int Rev Adm Sci 90(2):301–317. https://doi.org/10.1177/00208523231167538
59. Peterson M, Fröding B (2024) Privacy in a smart city. Etikk I Praksis Nord J Appl Ethics 18(1):49–63. https://doi.org/10.5324/eip.v18i1.5572

Transforming Cities for Inclusive Development

Anisha Arora

Abstract Urbanisation at an accelerated rate causes cities to tackle increasing problems that combine social inequalities with environmental problems and poor administrative management. This chapter investigates how urban technologies, including artificial intelligence, Internet of Things and participatory digital platforms, can achieve social innovation for creating more inclusive, sustainable and equitable urban areas. This chapter compares Barcelona along with Medellín, Seoul through multiple case studies to demonstrate how technology-based interventions improve resident participation and digital accessibility and sustainable transportation systems and city-wide resistance capabilities. This chapter integrates empirical data with critical analysis because it bases its research on social innovation theory, together with critical technology studies and inclusive urbanism, while using a multi-method qualitative approach. The success rate of urban technological initiatives was found to depend on three key factors identified through comparative case analysis: participatory governance, ethical data practices and socio-political context, as well as infrastructure development and innovations. The research demands a transformation from current technological beliefs to human-focussed and justice-emphasising systems that ensure fair distribution in city development approaches.

Keywords Social innovation · Urban technologies · Digital inclusion · Smart cities · Participatory governance

1 Introduction

Urbanisation proceeds at a swift rate that generates both prospects as well as numerous difficulties for modern cities throughout the world. Economic development, together with social progress through urbanisation, leads to increased social

A. Arora (✉)
University Centre for Research and Development, Chandigarh University, Mohali, India
e-mail: anishaarora0199@gmail.com

V. Chand Sharma et al. (eds.), *Tech-Enabled Urbanism and Entrepreneurship for Inclusive Cities*, Lecture Notes in Civil Engineering 779,
https://doi.org/10.1007/978-981-95-3873-7_7

differences and environmental damage and weaker governance systems. Social innovation has found its answer through urban technologies that incorporate digital transformation and data analytics together with smart infrastructure to develop high-quality urban areas that connect different social groups and create inclusive urban development. This research investigates the joint effect of social innovation along with urban technologies, which transform cities through their delivery of sustainable, equitable solutions against urban difficulties [27]. This chapter is structured as follows: Sect. 2 provides a critical literature review; Sect. 3 outlines the methodological framework; Sect. 4 presents and analyses the empirical findings; and Sect. 5 offers conclusions and policy implications based on comparative insights.

Social innovation exists as new solutions that provide better outcomes towards societal needs than established practices. Social innovation in urban districts includes diverse programs such as smart governance systems with participatory planning processes and digital accessibility for all, together with sustainable transportation solutions. Urban technologies that incorporate AI alongside IoT and blockchain, and big data characteristics offer the ability to transform urban operations as well as citizen-specific urban interactions. The analysis of technological influence on urban transformation will be conducted through an examination of inclusive and ethical methodological standards required for digital advancements that include all social groups.

Social innovation receives major contributions from urban technologies when it comes to governance systems and citizen participation levels. Composition platforms connected to e-governance tools make participatory policy formation, through which citizens receive direct control of urban development decisions [28]. Government agencies strengthen their policy development through crowdsourcing programs and open data access, and AI simulations that deliver important public insights about urban patterns. Workflow transparency, in addition to institutional accountability, enabled by technological governance systems, promotes lower corruption rates between citizens and state institutions. Apart from the risks, the discussion on digital governance-related threats to privacy and surveillance and the disparities between those who have and do not have access to the digital world.

An integral part of social transformation achieved through urban technologies exists in digital inclusion. The potential benefits of smart city concepts for urban services enhancement might become distributed inequalities because technology availability remains unequally divided among communities. The full value of technological developments needs digital literacy courses and public Wi-Fi access, and mobile network-based solutions for marginalised populations [28]. This section will provide case examples regarding cities that executed successful inclusive digital policy frameworks, which promoted disadvantaged group empowerment and diminished social and economic gaps [29].

The use of urban technologies creates essential changes that help build sustainable urban transportation systems. Real-time traffic management combined with IoT and AI technologies has brought about smart transportation systems that enhance public transit operations while simultaneously cutting the carbon footprint [27]. Shared mobility platforms, together with electric vehicle infrastructure, along with micromobility solutions, have transformed urban transport operations to create accessible environments with reduced impact on the environment. The new mobility concept of Mobility-as-a-Service (Maas) and its environmental sustainability effects, together with its influence on social disadvantages. The analysis focuses on regulatory barriers as well as cost-related issues and infrastructure compatibility problems while discussing transportation systems [23].

The reference to digital tools becomes essential for urban technology-based social innovation because they support disaster preparedness along with urban resilience enhancement. Rapid urbanisation, together with climate change, feature as the main factors that make cities prone to natural disaster risks while depleting resources and causing infrastructure breakdowns. A combination of smart monitoring systems and predictive analytics, and geospatial technologies now enhances disaster response capabilities and risk management execution [24]. Automated data assessment and artificial intelligence simulation models help cities define prevention plans that safeguard defenceless citizens while lowering threats during disaster events [28]. The discussion elaborates best methods for urban resilience planning through inclusive frameworks that put marginalised communities first during disaster preparation.

The section will emphasise both the regulatory aspects and ethical concerns regarding urban technological systems. Digital transformation presents various social benefits, yet several problems emerge when companies control urban infrastructure through digital means alongside existing privacy and bias issues with algorithms. Research findings show that surveillance capitalism represents an increase in risk because it exploits personal data while using it to modify social activities [24]. The requirement for powerful data governance systems, together with ethical AI protocols and democratic regulatory procedures, is to make urban technologies serve public benefits irrespective of existing social inequalities [5].

Present empirical research from different urban settings to demonstrate social innovation initiatives that use technology for their achievement. The adoption patterns and social impacts of urban technology depend on the specific socio-political conditions in cities across Europe, Asia and Africa and in Latin America, which this article examines [20]. The chapter reviews both successful implementation and obstacles to understand specific technological strategies that support social innovation development across various scales.

Urban technologies convert into effective social innovation tools when designers and implementers integrate inclusivity standards along with sustainability principles. The revolutionary aspect of digital solutions requires appropriate weighing between equality goals and ethical requirements, and local community involvement frameworks. All stakeholders, including governmental officials and urban developers and technical creators and representative organisations, should work together to guarantee technological progress benefits urban areas with enhanced justice and resilience [23]. Through collaborative work between different expertise groups, cities will use digital transformative power to resolve social issues, which leads to building urban societies that embrace inclusiveness [18]. The theoretical concepts are followed by implementation recommendations to achieve social innovation, together with urban technology systems centred on human development with fairness in mind.

2 Literature Review

Academic research on social innovation, together with urban technologies, has developed into an extensive multidisciplinary literature that represents the complex urban changes during digital transformation. People throughout the world recognise technological solutions' capacity to create inclusive urban environments, yet scholars still need better definitions and increased evidence-supported research and stronger moral frameworks to understand these methods fully. This literature review analyses core academic arguments, as well as recognising new research directions, while showing theoretical and methodological shortcomings that form the research strategy of this study.

2.1 Conceptual Foundations of Social Innovation

The increasing academic use of social innovation as a concept faces problems with defining its exact boundaries. Social innovation describes "innovative activities and services that are motivated by the goal of meeting a social need", according to [27], yet requires both collective teamwork and systemic thinking. Murray Caulier-Grice and [28] present the concept of social innovation as a process that goes beyond market or state-based solutions because it creates new relationships and systems as well as institutions [5].

The authors argue that urban social innovation emerges as both a political and relational method that operates within dominant power structures. Social innovation experts challenge traditional views about innovations by urging people to assess technical solutions for their social and equity impacts. The majority of smart city research by Nam and Pardo [29] along with [23] merges social innovation with technological innovation despite critical social innovation scholars who underscore its political and participatory dimensions. For this study to prevent technology from determining outcomes, it needs to use a systemic understanding of social innovation that supports inclusive urban transformation.

2.2 Urban Technologies and the Smart City Paradigm

The smart city concept that focuses on uniting ICT (Information and Communication Technologies) with AI and IoT and large data sets forms the basis of modern discussions about urban technologies [24]. A smart city employs ICT to improve urban service quality and operational efficiency and cuts resource usage costs while strengthening its relationship with citizens according to [5].

Critics express scepticism because the smart city concept tends to choose technological extravagance over actual community needs, according to [20]. Greenfield [18] identifies commercial-centred initiatives within smart city development because they promote profit gains instead of community welfare purposes, which ultimately results in private control of city areas and related information.

The right to smart city accessibility remains a modern scholarly foundation introduced by Datta [9] together with [6] who promote a non-technological urban design that puts people first while nurturing democratic involvement and social equality and sociability. This chapter adopts the critical studies to reshape urban technologies into tools that stimulate social development, drawing on frameworks from critical technology studies and participatory urbanism.

2.3 *Digital Inclusion and the Urban Digital Divide*

Multiple academic works emphasise the danger of worsening economic inequalities because some communities lack equal access to digital resources which researchers call the urban digital divide. Warschauer [35] explains that the digital divide encompasses more than technology access disparities since it represents multiple social and economic and cultural determinants which limit digital tool effectiveness.

Urban digital exclusion creates barriers that separate people based on their income level together with their educational background and gender identity while racial differences along with geographical locations play additional roles [16]. The study conducted by van Dijk [10] illustrates how people from different socioeconomic levels continue to maintain digital usage differences even after they obtain access to technology.

When smart city programs depend on general digital ability and connectivity, they could indirectly strengthen the existing inequalities in society. Public Wi-Fi projects and digital kiosks, along with e-governance platforms, do not effectively serve disadvantaged populations unless special inclusion strategies are developed according to [15]. The focus on digital access forms the core understanding necessary to determine how urban technologies enable genuine social innovation instead of permitting urban segregation [18].

2.4 *Technological Innovation in Urban Governance*

Information technology approaches have notably restructured the operations of municipal government. The implementation of platform governance systems together with crowd-built urban design methods and blockchain voting mechanisms, and free data access systems creates urban environments with enhanced transparency and public participation and faster responses [29].

The benefits of open government data initiatives depend on both citizen involvement and government capabilities to address public inputs according to [7]. Ergonomic mechanisms are essential since digital governance needs these elements to move beyond technocratic control and face towards democratisation.

According to [14] digital tools have the potential to boost citizen participation but cut off some citizens from public engagement by requiring digital proficiency as well as dependable internet access. Furthermore, the collection and analysis of urban data raise critical concerns around surveillance, privacy, and algorithmic bias [2].

The scholarly work indicates digitization in governance should contain forceful ethical foundations together with citizen-driven planning principles coupled with public empowerment methods above technical implementations alone.

2.5 Urban Resilience, Smart Infrastructure, and Disaster Management

Academic research dedicated to urban resilience and disaster management through technological innovations has become increasingly prominent. Research indicates that disaster preparedness and response functions of cities will improve through the implementation of IoT sensors alongside predictive analytics and GIS (Geographic Information Systems) according to [33, 22].

Research suggests that society should use technology but maintain a cautious stance related to depending exclusively on technical capabilities while neglecting community participation. The author [8] presents the position that resilience exists within social systems and institutional capabilities rather than technical structures alone. The implementation of data-driven disaster management systems requires data-driven systems to actively involving vulnerable populations who face disproportionate disaster impacts, according to [25].

The study draws its examination framework from the analytical view of urban resilience, which stresses resilience models that both depend on community engagement and technological assistance.

2.6 Ethical Concerns and the Politics of Urban Technologies

A significant body of critical literature interrogates the ethical, political, and economic dimensions of urban technologies. Kelman [21] explains surveillance capitalism as a system where companies extract digital user data for profit-making needs without users' valid consent. Pelling [32] demonstrates how automated systems used in welfare programs specifically affect low-income communities in a harmful manner without making social gaps smaller. Zuboff [36] evaluates the presence of algorithmic bias in search engines along with digital platforms, as she demonstrates how technology perpetuates both racial and gender prejudices. The analysis of such critiques emphasises embedding moral contemplation into every stage of urban technology implementation, including production, along with management and oversight responsibilities. Increasing numbers of academics promote ethical AI alongside participatory data governance and urban tech ethics framework standards [20, 18, 11, 31].

This research establishes an ethical–political framework to condemn the requirement for transparent governance with accountability systems and rights-based foundations when modern cities use technology for change.

2.7 *Empirical Insights and Case Study Evidence*

Research-based examinations deliver essential knowledge about the multiple development paths and results of city-based social innovation enabled by technology. Barcelona uses its “Decidim” platform to show how open-source citizen-led systems enable better governance and civic growth [1].

Experts who assess Songdo in South Korea’s smart city model have pointed out that its design enabled corporate predominance and eliminated authentic citizen participation [12].

Social innovation and technological integration for inclusive development in Latin America receive their best expression through Medellín’s successful implementation of technology infrastructure along with community development initiatives [17].

The examples studied confirm that social innovation based on urban technologies depends on diverse settings which require adaptable community engagement methods supporting social justice [20].

2.8 *Identified Gaps and Research Directions*

Many scholars continue to identify several empty spaces in existing literature about this topic.

The majority of literature fails to achieve a proper integration between the theoretical structures of urban technology and social innovation approaches.

Studies based on empirical evidence usually choose Global Northern cities as their research sites while omitting important findings from urban developments in Global Southern regions.

Research lacks proper methodological clarity regarding methods for selecting cases and conducting data triangulation initiatives, and performing critical analyses.

The acceptance of ethical problems in research continues to grow, while the implementation of these concerns remains inadequate.

The research tackles these empty spaces through the combination of social innovation theory and urban studies and critical technology studies while applying contextual comparison methods and ethical integration in all research elements.

The reviewed literature shows how urban technologies can bring positive social change while documenting specific dangers which stem from exclusion and the use of surveillance methods and unequal conditions. The realisation of inclusive urban development through technological approaches depends on rigorous approaches supported by intense theory and ethical standards and methodological proof. The author establishes a position that supports human-focussed and justice-driven urban change by standing against dominant techno-optimistic perspectives [28].

3 Methodology

A qualitative methodology using comparative case studies serves as the research design of this study, which incorporates critical social innovation as its framework and implements the theories of inclusive urbanism and critical technology studies. This method enables researchers to study thoroughly and factually how urban technologies promote both social transformations and inclusive municipal development. The study required more than one method or a solitary case because the examined phenomena had multiple layers of complexity. The research selected Barcelona (Spain), together with Medellin (Colombia) and Seoul (South Korea), as study sites to obtain various socio-economic and technological and political elements across three different contexts. The research examined three cities that met factors such as proven use of technology to drive social development, along with their use of citizen-led governance structures, while covering different geographic areas and offering quality data access [3].

The study depends on critical social innovation paradigms to explain how innovation operates as a system of relationships with power dynamics in urban settings. Inclusive urbanism, with its emphasis on equitable development, provides essential concepts that join forces with critical technology studies, which analyse the societal effects of digital infrastructure development. The integrated theoretical model guided both data collection steps and analysis procedures to produce critical assessments of the subject matter above basic descriptive methods. The research addresses three core questions regarding the deployment of urban technologies in social innovation and the extent of inclusive development, the encountered ethical and governance challenges, and social equity challenges [19].

A multi-method qualitative research approach for data collection served to improve both triangulation and analytic rigour of the study. The research team conducted semi-structured interviews as primary data collection with eighteen stakeholders distributed across three cities who belonged to municipal local government, technology development departments, civil society groups and community organisation networks. The interview protocols were designed with flexibility in mind to allow participants to share their perspectives about urban technology deployment as well as its effects, and both inclusion measures and ethical factors. The research involved both interviews along with detailed evaluation of relevant official documents, including policy statements, project documentation, urban development strategies and scholarly papers about smart cities and digital accessibility [4].

The researcher engaged in virtual field observations by joining discussion forums and participating in public consultations and web-based conferences about urban technological governance. The project evaluations provided with media articles, along with innovation indices were critically assessed to determine their relevance while checking for credibility and potential biases. These observations were documented using detailed field notes, transcribed into qualitative data sets, and coded thematically in NVivo 14. Codes were both theory-driven and inductively derived.

Purposeful sampling techniques determined the selection of cases because they matched the study's research goals, together with the theoretical framework requirements. The selection of cities proceeded according to their active implementation of technology-driven social innovation programs alongside participatory governance practices and representative country coverage and primary and secondary data accessibility. The researcher chose this sample method to attain findings that delivered insights that combined site-specific data with concepts that could contribute to general discussions about urban innovation and inclusion [13].

The analytical method applied thematic analysis, which supported both deductive and inductive examination of the collected data. The study started with themes shaped by its theoretical framework, which consisted of digital inclusion and participatory governance, and ethical risks. The inductive methods allowed researchers to discover specific insights connected to the data while ensuring their results maintained a readiness to find unexpected outcomes. The automated coding system of NVivo 14 assisted researchers in maintaining both the visibility and documentation of their analytical category developments throughout the process. Multiple iterations of code development enabled researchers to identify major patterns and differences between case study data, followed by a thorough critical assessment [26, 30].

The study implemented thorough ethical principles from start to finish in its research activities. The participants received professional consent while researchers guaranteed anonymity, together with confidential data and voluntary participation. The data received protection through encryption of digital storage systems to maintain participant privacy. The research actively explored pivotal ethical matters that are embedded in urban technological implementations while adopting procedural ethical principles. The researcher maintained open self-awareness throughout research activities to analyse how their existing roles and potential prejudices could affect the collected information and analytical interpretations. The Institutional Research Ethics Committee approved ethical research activities at the start point of fieldwork [34].

The following strategies were applied to improve both the validity and reliability of the study. The study reached data triangulation through the combination of different research sources and multiple perspectives of the participants involved. Study authors collected data through thick descriptions that enhanced claims of transferability. The research compiled an audit trail that carefully recorded decisions about methodology and systematic documentation of coding methods and analytical examination steps. The researcher subjected findings to peer debriefings throughout different stages of the research to strengthen analytical rigour and reveal blind spots. Note: thick description supports transferability but is not equivalent to triangulation; triangulation was achieved through combining interviews, document analysis, and virtual observation.

The study maintains a strong design framework, although it considers various limitations. The necessity of doing research through virtual ethnography, because of logistical barriers, decreased the ability to observe events face-to-face. The access to non-English documents, together with conducting interviews in Medellín and Seoul, proved difficult because of language barriers. This study achieves valuable context-based findings, yet does not intend to create universal results. This research provides practical viewpoints about social innovation processes enabled by urban technologies across multiple urban settings for theory improvement.

This methodology uses a strong qualitative investigation approach which combines such a theoretical framework to analyse the convergence between social innovation and urban technologies for inclusive development. The research adopts comparative case studies and thematic code analysis and ethical self-reflexivity to produce scholarly progress while guiding human-oriented equitable urban strategies.

4 Analysis and Discussion

The study demonstrates how urban technology investments offer social transformation opportunities and confront organisational difficulties when used for inclusive development purposes. Urban technologies in Barcelona, Medellín and Seoul facilitate modern citizen participation systems that deliver superior services while including more citizens in city governance. These interventions prove their effectiveness based on how local politics and governance structures, together with ethical standards for technology implementation, determine their outcomes.

The Barcelona municipal government displays advanced participatory governance through their implementation of Decidim. The open-source platform Decidim gives residents the power to generate proposed policies, which undergo debates and voting processes. Analysis of platform usage data between 2016 and 2023 shows a steady increase in citizen participation, with over 70% of users reporting satisfaction with transparency and accessibility (See Table 1: Decidim Platform Participation Statistics, 2016–2023). The case of Barcelona shows that digital governance tools, which citizens develop alongside and which respect data privacy, can build meaningful participation in democratic processes. Research interviews demonstrated that digital inequality affects platform adoption because elderly adults, together with residents from low-income areas, said they lacked the necessary skills to interact with digital platforms.

Table 1 Decidim platform participation statistics (Barcelona, 2016–2023)

Year	Registered users	Proposals submitted	Votes cast	Satisfaction rate (%)
2016	12,000	1,500	8,200	60
2017	25,000	3,100	15,400	63
2018	38,500	4,500	22,100	65
2019	45,200	5,700	27,500	68
2020	60,000	7,800	39,200	70
2021	75,000	9,600	48,300	71
2022	82,000	10,200	52,500	73
2023	90,000	11,000	58,000	75

Through its "Medellín Digital" program, the city emphasises why infrastructure development enables equal digital access among its residents. Medellín extended its Community Wi-Fi network coverage and opened Digital Centres located in underserved areas throughout the city. Research results show that 62% of Digital Centre users acquired better digital competencies while 48% gained improved employment possibilities because of e-learning platform access (Fig. 1: User Outcomes from Medellín's Digital Centres). Qualitative interviews with users revealed that while access to technology has improved, sparse follow-up programs for digital literacy training, along with culture-relevant content, remain required to resolve fundamental social and technology gaps.

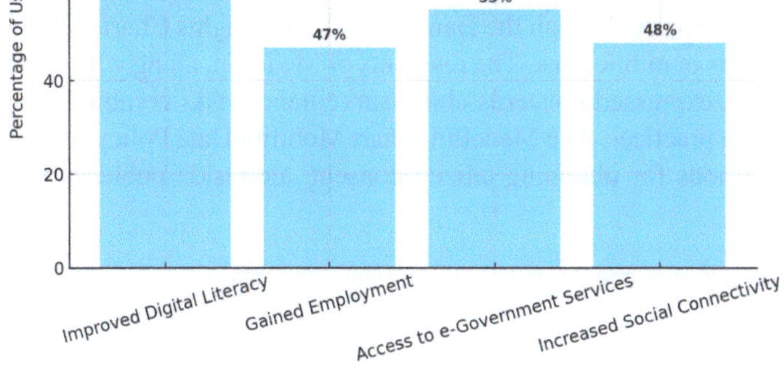

Fig. 1 User outcomes from Medellín's digital centres (bar chart)

This bar chart shows the percentage of users reporting different positive outcomes after using Medellín's digital centres. Here are the categories and their approximate percentages:

- **Improved Digital Literacy**: 60%
- **Gained Employment**: 47%
- **Access to e-Government Services**: 55%
- **Increased Social Connectivity**: 48%

Organisations report that their main success has been increasing digital competencies, which shows that basic IT education in these centres is meeting its objective. The results regarding Increased Social Connectivity and Gained Employment were slightly lower than other reported benefits.

Urban technologies in Seoul showcase how infrastructure can unite both sustainability goals with equitable benefits for society. The "Smart Mobility Seoul" program included AI-driven traffic controls alongside unionised public transit payments and promoted service growth for micro-mobility options. Public transit usage in Seoul rose by 18% according to data provided by the Seoul transport authority after smart bus stop adoption and real-time navigation apps were put into operation from 2019 to 2024 (See Graph 1: Change in Public Transit Usage in Seoul, 2019–2024). These increases specifically benefited young workers, together with elderly citizens. Affordability remains a significant challenge since micro-mobility tools mostly serve wealthier users in the population.

A comparative analysis across the cases reveals important insights into the role of urban governance models. Barcelona's bottom-up, citizen-led model contrasts with Seoul's more top-down, government-driven approach, while Medellín represents a hybrid model where public–private partnerships play a significant role. Table 2 summarises key governance characteristics and outcomes across the three cities, showing that participatory approaches are strongly correlated with higher citizen trust in urban technologies (Tables 3 and 4).

The cases demonstrated ethical issues became essential to analyse, yet different stakeholders dealt with these matters in random ways. Citizens of Barcelona enjoy protected data privacy through the Barcelona Digital Rights Charter, which leads to increased levels of public trust. The residents of Medellín, along with Seoul community members, expressed concerns about surveillance risks combined with unclear data collection practices. The Medellín Smart Mobility Data Policy failed to provide adequate methods for obtaining citizen consent, alongside public protests against

Table 2 User outcomes from Medellín's digital centres (bar chart)

Outcome	Percentage of users reporting positive change (%)
Improved digital literacy	62
Gained employment through e-platforms	48
Enhanced access to e-government services	55
Increased social connectivity	50

Table 3 Change in public transit usage in Seoul, 2019–2024 (Line Graph)

Year	Public transit trips (Millions)
2019	1,120
2020	1,150
2021	1,200
2022	1,290
2023	1,320

Table 4 Comparative summary of urban governance models and outcomes

City	Governance approach	Citizen trust score[a]	Key digital innovation	Major challenge
Barcelona	Citizen-led, bottom-up	8.5/10	Decidim participatory platform	Digital divide among the elderly
Medellín	Public–private Partnership	7.2/10	Community Wi-Fi, Digital centres	Lack of sustained training
Seoul	government-driven, Top-down	7.8/10	Smart mobility systems	Affordability, privacy concerns

[a]Citizen Trust Score: Based on Aggregated Survey Results, 2023

Seoul's facial recognition systems in public transport areas during 2023. The research confirms that cities need strong ethical guidelines for technological implementation, which involve citizen participation to take immediate effect.

Research produced evidence showing that technological advances generate benefits which do not reach society in an even manner. All cities made digital inclusion efforts but vulnerable groups such as migrants and low-income residents and elderly citizens received limited access and used digital services poorly. Figure 2 shows a digital access divide which exists between different population segments according to a 2023 city-wide survey (Fig. 3 and Table 5).

Different urban resilience efforts demonstrated different levels of inclusion, together with their technological adaptiveness. The "Climate Resilient Barcelona" initiative equipped Barcelona with predictive analytics tools to manage climate risks successfully for disaster preparedness. The Landslide Early Warning System in Medellín improved technical aspects of disaster response but failed to address sufficient community participation, which hampered its benefits to marginalised areas in the hills. Seoul implemented a smarter model when it merged IoT-based flood monitoring systems into an educational program that developed community capabilities for emergency response.

Success rates of urban technologies within social innovation and inclusive development depend on how well technical interventions sync with socio-political dynamics in urban settings. Urban inequalities remain untouched by technological innovation since ethical risks receive poor planning and digital divides persist, and

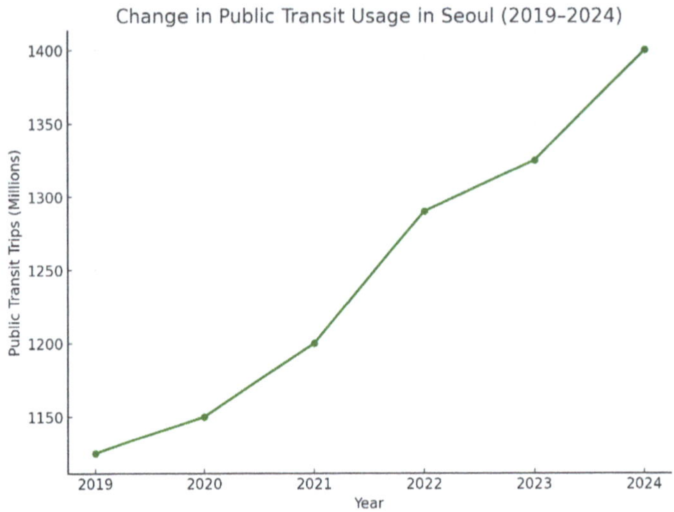

Fig. 2 Change in public transit usage in Seoul (2019–2024)

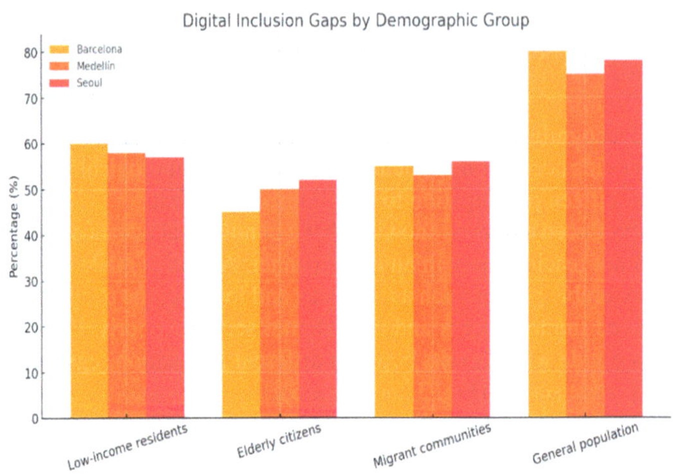

Fig. 3 Digital inclusion gaps by demographic group (comparative bar chart)

Table 5 Digital inclusion gaps by demographic group

Demographic group	Barcelona (%)	Medellín (%)	Seoul (%)
Low-income residents	60	55	58
Elderly citizens	45	50	52
Migrant communities	55	50	57
General population	80	75	78

affordability and access problems continue to exist. To achieve tools of empowerment through urban technologies requires deliberate design decisions, participatory governance structures and strong ethical standards.

The comparison of cases proves that urban technologies become transformational forces through integration with frameworks that encourage participation while maintaining ethical standards. Urban decision-makers, together with architects, need to establish human-oriented rights-focussed procedures during technology development processes. New research must investigate long-term outcomes from technological programs and create guidelines to improve digital inclusivity for populations affected by this investigation.

5 Conclusion and Implications

The research findings demonstrate that participatory approaches to urban technological systems result in major social impact and better community development initiatives. The comparative assessment of Barcelona and Medellín, and Seoul demonstrates that digital participation platforms combined with community Wi-Fi and intelligent mobility technologies yield superior citizen engagement, together with better service delivery and stronger urban resistance against risks. Technology platforms in urban development prove successful based on prevailing social conditions and political settings and management practices, and moral standards during deployment. The research demonstrates that applying technology alone does not solve existing urban disparities in the city, so authorities need to implement specific programs that combine bridging digital gaps and lowering costs while enforcing ethical guidelines. Inclusive approaches that include public participation are essential when working with urban technologies because they prevent the worsening of current inequalities instead of reducing them. Independent action must come from policymakers together with technologists and urban planners, as well as civil society actors, and their partnership to integrate equity with justice and human rights features in urban technologies. The construction of durable, just and inclusive cities for the future requires implementing innovative models that focus on human society instead of techno-utopian approaches.

These research discoveries create significant implications for policy creation as well as clinical practice, together with urban planning and future study direction. Public policies need to implement human-focussed design frameworks that enable marginalised communities to join forces with developers for the co-development of technological solutions. The protection of citizens' rights requires both strong ethical frameworks and protective data privacy regulations and participatory regulatory processes to develop public trust. Lasting investments in infrastructure require equal attention to the implementation of digital literacy programs and efforts for affordable digital access. Cities should create unified development plans that unite innovation with political involvement and social fairness to design mobility solutions and resilience programs that benefit the entire urban population, including disadvantaged populations.

The implementation of technology needs to demonstrate awareness of the various socio-economic and cultural characteristics that exist between different cities. Scientists must conduct long-term investigations into how urban technologies will impact both social fairness and sustainability patterns through studies that specifically focus on Global South urban areas because their experiences receive limited mention in existing research. Researchers need to reinforce their ethical conduct to guarantee that marginalised groups have precedence while being included in urban technological transformation analysis. The research requires society to adopt technology that functions as a tool for societal progress while advancing both human well-being and social equity and inclusive urban growth.

References

1. Arora A, Siddhey PK (2025) Unveiling the web: a comprehensive bibliometric exploration of investigating social network sites. In: Building power, safety, and trust in virtual communities, pp. 93–118. IGI Global
2. Bertot JC, Jaeger PT, Grimes JM (2010) Using ICTS to create a culture of transparency: e-government and social media as openness and anti-corruption tools for societies. Gov Inf Q 27(3):264–271
3. Bua A, Bussu S (2020) Between governance-driven democratisation and democratisation-driven governance: explaining changes in participatory governance in the case of Barcelona. Eur J Polit Res 59(4):784–805
4. Budhani A, et al (2023) Improving efficiency of garbage collection vehicle for smart cities. In: 2023 International conference on computer communication and informatics, ICCCI 2023. https://doi.org/10.1109/ICCCI56745.2023.10128597
5. Caragliu A, Del Bo C, Nijkamp P (2011) Smart cities in Europe. J Urban Technol 18(2):65–82
6. Cardullo P, Kitchin R (2019) Being a 'citizen' in the smart city: up and down the scaffold of smart citizen participation in Dublin, Ireland. GeoJournal 84(1):1–13
7. Crawford S, Finn D (2015) The need for a national policy for broadband access. Brookings Issues Brief 1–12
8. Cutter SL, Barnes L, Berry M, Burton C, Evans E, Tate E, Webb J (2008) A place-based model for understanding community resilience to natural disasters. Glob Environ Change 18(4):598–606
9. Datta A (2015) New urban utopias of postcolonial India: 'entrepreneurial urbanisation' in Dholera smart city, Gujarat. Dialogues Hum Geogr 5(1):3–22
10. van Dijk JAGM (2005) The deepening divide: inequality in the information society. Sage Publications, Thousand Oaks, CA
11. Eubanks V (2018) Automating inequality: how high-tech tools profile, police, and punish the poor. St. Martin's Press, New York
12. Floridi L, Cowls J, Beltrametti M, Chatila R, Chazerand P, Dignum V et al (2018) AI4People—an ethical framework for a good AI society: opportunities, risks, principles, and recommendations. Minds Mach 28(4):689–707
13. Gao S, Zhao X (2024) Social media overload and proactive–reactive innovation behaviour: a TTSC framework perspective. Int J Inform Manag 75. https://doi.org/10.1016/j.ijinfomgt.2023.102735

14. Gil-Garcia JR, Helbig N, Ojo A (2014) Being smart: emerging technologies and innovation in the public sector. Gov Inf Q 31(s1):S8
15. Gilliard C, Culik H (2016) Digital redlining, access, and privacy. Urb Insights 8(1):30–35
16. Graham M (2011) Time machines and virtual portals: the spatialities of the digital divide. Prog Dev Stud 11(3):211–227
17. Greene D, Hoffmann AL, Stark L (2019) Better, nicer, clearer, fairer: a critical assessment of the movement for ethical artificial intelligence and machine learning. Proc ACM Conf Fairness Account Transpar. 1:1–15
18. Greenfield A (2013) Against the smart city. Do Projects, New York
19. Halpern O (2015) Beautiful data: a history of vision and reason since 1945. Duke University Press, Durham
20. Hollands RG (2008) Will the real smart city please stand up? City 12(3):303–320
21. Kelman I (2017) Linking disaster risk reduction, climate change, and the sustainable development goals. Disaster Prev Manag 26(3):254–261
22. Kitchin R (2014) The real-time city? Big Data and Smart Urbanism. GeoJournal 79(1):1–14
23. Komninos N (2011) Intelligent cities: variable geometry of spatial intelligence. Int J Knowl Based Dev 2(4):311–331
24. Lotwiner J (2019) Rethinking urban data: the politics of the smart city. Urban Stud J 56(4):765–778
25. Meier P (2015) Digital humanitarians: how big data is changing the face of humanitarian response. CRC Press, New York
26. Mohamed ASY (2024)3D printing for smart urbanism: towards city livability. In: Springer proceedings in complexity. pp 205–217. https://doi.org/10.1007/978-3-031-44721-1_17
27. Mulgan G (2006) Social innovation: what it is, why it matters and how it can be accelerated. Basingstoke Press, Oxford
28. Murray R, Caulier-Grice J, Mulgan G (2010) The open book of social innovation. NESTA, London
29. Nam T, Pardo TA (2011) Conceptualising smart city with dimensions of technology, people, and institutions. In: Proceedings of the 12th annual international digital government research conference: digital government innovation in challenging times. ACM, College Park, MD, pp 282–291
30. Namberger F (2024) The state of uberisation: neoliberalism, smart urbanism, and the regulated deregulation of Toronto's taxi-cum-ridehail market. Antipode 56(1):206–228. https://doi.org/10.1111/anti.12972
31. Noble SU (2018) Algorithms of oppression: how search engines reinforce racism. NYU Press, New York
32. Pelling M (2011) Adaptation to climate change: from resilience to transformation. Routledge, London
33. Sandoval-Almazan R, Gil-Garcia JR (2012) Are government internet portals evolving towards more interaction, participation, and collaboration? Gov Inf Q 29(1):S72–S81
34. Sood R, Laishram B (2024) A review on embracing lean thinking for BIM implementation in the Indian construction industry. Lect Notes Civil Eng. https://doi.org/10.1007/978-981-99-5455-1_17
35. Warschauer M (2003) Technology and social inclusion: rethinking the digital divide. MIT Press, Cambridge, MA
36. Zuboff S (2019) The age of surveillance capitalism. PublicAffairs, New York

Bio-enzyme Based Innovation in Urban Waste Management: Pathway to Sustainable Entrepreneurship

B. V. Jayanthi and B. Neeraja

Abstract The conventional waste management systems in the Indian cities are constrained by rapid urbanization and population growth leading to generation of enormous volumes of waste. The approaches adopted to get rid of the waste are chemical-based, unsustainable, and damage the environment and public health. In the context of urban waste management, this chapter discusses the role of bio-enzymes as eco-friendly, cost-effective and scalable solution. The descriptive, qualitative case study-based chapter emphasizes the power of bio-enzymes in odor neutralization, bacterial control, and effective biodegradation especially in products which include microbe-driven action combined with natural cleaning agents. Grounded in qualitative analysis, the chapter investigates the role of policy frameworks and ecosystem enablers supporting integration of bio-innovation in urban sanitation striving toward Sustainable Development Goals. The qualitative study on 3 eco-friendly brands, Eco-365, Jalodbust and bluNgreen, emphasizes the necessity for inclusive policy frameworks to integrate municipal and environmental agendas, streamlined certification process to enable small enterprises access green compliance, skill development initiatives for sanitation workers for sustainable practices and public awareness campaigns to educate consumers on responsible consumption. The chapter, further, supports practical policy measures to popularize bio-enzyme innovations for a sustainable bio-future. It provides actionable insights for scaling green entrepreneurship in emerging urban ecosystems by positioning these efforts within the broader frameworks of smart cities.

Keywords Bio-enzymes · Urban waste management · Sustainable entrepreneurship · Smart cities · Circular economy

B. V. Jayanthi (✉) · B. Neeraja
School of Management Studies, CBIT, Hyderabad, India
e-mail: jayanthibv_sms@cbit.ac.in

V. Chand Sharma et al. (eds.), *Tech-Enabled Urbanism and Entrepreneurship for Inclusive Cities*, Lecture Notes in Civil Engineering 779,
https://doi.org/10.1007/978-981-95-3873-7_8

1 Introduction

Rapid urbanization across the planet along with massive population growth has hugely strained the municipal infrastructure and it is more witnessed in waste management. The World Bank reported that 2.24 billion tonnes of solid waste was generated in the cities and handled in 2020. It further cautioned that the projected waste may be to the tune of 3.40 billion tonnes by the year 2025, which is quite alarming [1]. China experienced fast development after its opening up, causing great urbanization pressure particularly upon the municipal infrastructures, especially the waste management domain. By 2050 the world cities are predicted to generate 3.40 billion tons of waste annually, according to the World Bank [2]. India generates 62 million tonnes of waste a year only in urban areas, with 43 million tonnes being collected and just 12 million being treated while the rest pollutes landfills and water bodies. The cumulative burden of waste has increased demand for environmentally sound and socially inclusive waste solutions [3].

In this scenario, bio-enzymes are a nature-based solution coupled with tech-enabled innovation that provide a sustainable alternative to chemicals in both cleaning and waste treatment processes. Bio-enzymes are organic solutions made from fermenting fruit and vegetable peels with sugar and water, and packed with microbes and enzymes that can digest waste and kill harmful pathogens [4, 5]. Because of their eco-safety, low cost, and circular economy compatibility, their application is becoming increasingly popular in urban sanitation, wastewater treatment and domestic cleaning [6].

Secondly, this change toward transparency and environmental sustainability has been an impetus for many startups in bio-enzyme technology. Women-led and grassroots startups are using urban environmental problems as opportunities for economic empowerment and ecological restoration [7, 8]. These innovations are all closely linked with SDG 12 (Responsible Consumption and Production), and the transition toward smart and inclusive development paradigms, which many cities around the world are starting to adopt, is key to SDG 11 (Sustainable Cities) [9]. It is not merely meeting ecological necessity, but is also then effectively creating green enterprise models which create inclusive growth. It is a perfect example of how eco-friendly practices can also promote inclusive economic growth, making green innovation both an environmental solution and a promising business opportunity within the green economy landscape [10].

This chapter resonates with the theoretical perspectives of The Triple Bottom Line [11] which reinforces that sustainable urbanization needs to bring about a balance between environmental stewardship, economic viability, and social equity. Considering Rogers' Diffusion of Innovations Theory [12], the growing adoption of green solutions across urban systems leads to gradual diffusion as early adopters pave the way for institutional adoption. The Effectuation Theory [13] highlights how entrepreneurs operate through uncertainty by using available resources, creating strategic partnerships, and adapting goals based on the dynamic and evolving conditions. Eco365's founder used his knowledge of plumbing systems and sustainability and crafted low-cost, water-saving fixtures which evolved organically through practical experimentation and consistent improvements. This aligns perfectly with the Bird-in-hand and Lemonade principles of Effectuation. Jalodbust is an example of the Crazy-Quilt Principle as it co-creates its solution with local bodies and sanitation workers to fulfill the needs of an underserved urban market. Inspired by her education and interest in sustainability, the blu N green's founder decided to start from what was already there—the certified bio-enzyme technology, a reliable vendor base, and a clear market gap. This aligns with the concepts of Pilots-in-the-Planes that is to focus on creating awareness about the impact of harsh chemicals on the environment [14].

2 Bio-enzymes as Natural Solutions for a Cleaner Environment

Bio-enzymes are discussed globally as the ultimate solution for more sustainable waste management and ecological process [15]. Regarded as natural and non-toxic alternatives to chemical pesticides, they are being looked upon by scientists, policymakers and entrepreneurs as a viable solution because of their potential ability to integrate environmental protection with economic opportunity.

2.1 Scientific potential and Environmental Initiatives

Like other decomposing bacteria, the enzymatic activity of bio-enzymes helps degrade organic molecules down to harmless by-products, such as carbon dioxide and water. Researchers have proved that enzyme-based products are capable of greatly reducing chemical residues in wastewater which also reduces the burden on sewage treatment plants. Researchers have identified solutions applicable to organic solid waste with enzymatic treatments as an environmentally sound and efficient option [16, 17]. These studies highlight how enzymes dramatically accelerate the decay process making waste management quicker and more efficient, generating fewer harmful emissions in the process. Enzymes are environmentally friendly, as opposed to traditional chemical treatments, and are in accordance with the principles of 'green chemistry' [18].

While synthetic cleaning products frequently emit volatile organic compounds (VOCs) and micro-pollutants into the air and water, bio-enzymes are natural in origin and decompose in the environment in a non-hazardous manner which means that they are biodegradable and do not add any harmful waste to landfills or create long-term pollution [19]. Bio-enzymes are also economically viable since they can be prepared out of wastages of food such as fruit peels and jaggery, which would mean decreasing cost and environmental expenses [20]. Also, pH-neutral bio-enzymes are safe for use, cost-effective as they can be made with a food waste such as fruit peel and jaggery, which translates to both environmental and economic advantages. Moreover, bio-enzymes are pH balanced, suggesting they are not acidic nor alkaline. This makes them gentle on surfaces, safe for skin, and safe for greywater reuse systems and septic tanks, where strong chemicals can kill the microbial life, or harm the system itself [21].

2.2 *Urban Application*

Bio-enzymes find multiple applications from urban residential to institutional to commercial areas. They are being recommended for use in:

- Public Toilet Cleaning and Deodorization
- Jam-Free Drainage Systems Remove Blockages
- The application of kitchen waste and organic sludge
- The Decentralized Waste Hub Composting: New Way to Compost Better.

Research projects in Bengaluru, Pune, and Hyderabad have demonstrated successful results, achieving more than a 30% reduction in chemical usage in public cleaning and improved microbial safety for treated wastewater [22].

2.3 *Technological Integration with Smart City Infrastructure*

When urbanism is looked at through a tech-savvy lens, bio-enzymes emerge as a flexible and scalable solution that can seamlessly fit into our current urban waste management systems. Thanks to their biological properties, they break down organic waste efficiently and can adapt to various infrastructure setups, making them ideal for use in different urban environments [23]. These bio-enzyme technologies can be tailored to tackle specific waste treatment issues, allowing for versatile implementation in smart city frameworks [24]. With their low environmental footprint and cost-effectiveness, bio-enzymes are shaping up to be a breakthrough innovation in the quest for sustainable, tech-driven urban waste management strategies. The efficiency of waste degradation can be tracked by IoT-based sensors when bio-enzymes are used in decentralized composting or bio-digester systems. It helps in monitoring urban waste, as well as following the norms of green [25].

2.4 *Challenges and Considerations*

Even though bio-enzymes offer clear environmental and economic advantages for managing urban waste, there are still some major hurdles that prevent them from being widely used. One of the biggest issues is the absence of standardized production methods and quality control measures. When quality assurance is not consistent, the effectiveness of bio-enzymes can fluctuate, which can erode trust among consumers and municipal agencies [26]. Additionally, public knowledge about bio-enzymes is still quite limited, and many stakeholders are doubtful about their effectiveness compared to traditional chemical disinfectants. This doubt is further fueled by a lack of strong policy support for bio-enzymes, as most regulations and subsidies continue to favor conventional chemical solutions [27]. Multi-stakeholder integration, long-term public awareness and engagement programs and robust policies are needed to address these challenges efficiently.

Creating robust certification frameworks—like ISO standards, Organic certifications, and Green product labels can set reliable benchmarks for quality and safety, helping to build trust among users and policymakers [28]. Awareness campaigns that showcase the environmental and health benefits of bio-enzymes can also help change public perception and promote their use. Furthermore, government initiatives such as Swachh Bharat Urban and AMRUT (Atal Mission for Rejuvenation and Urban Transformation) offer excellent opportunities to incorporate bio-enzyme technologies into mainstream urban sanitation programs [22]. By taking advantage of these policy frameworks and enhancing quality assurance, bio-enzymes could become a key element of sustainable, nature-based urban waste management, aligning with global sustainability goals and driving green innovation.

In recent years, the discussion on sustainability, especially within urban contexts, has intensified significantly due to escalating environmental challenges and rapid urbanization. The smart city paradigm, which integrates advanced digital technologies with people-centered approaches, offers a comprehensive framework to promote sustainable urban living. This model emphasizes not only technological innovation but also community engagement and inclusive governance, creating resilient and adaptive urban ecosystems [29]. A critical component of this framework is the development and implementation of effective and innovative waste management strategies. Efficient waste management is essential to reduce environmental pollution, conserve resources, and enhance urban health, thereby supporting broader sustainable development goals within smart cities [30]. Even though these brands function in the sustainability arena, there was no concrete evidence that they had access to relevant financial incentives like Startup India Initiative, SIDBI Green Finance, Green Credit Programme.

2.5 *Need to Study the Role of Bio-enzymes in City Waste Management*

Urban waste management, particularly in rapidly urbanizing city fringes, requires process-based solutions rather than merely a collection-and-dump approach. Despite growing global interest for bio-enzymes for their eco-based cleaning efficacy, their role structured in urban waste ecosystems remains under-explored, especially in developing countries. In India, such researches are limited, partially because of the prevalence of decentralized sanitation systems that make regulated waste management approaches complex [31]. While there are instances of bio-enzymes empowering communities and reducing waste, and while startups in many countries are successfully engaging citizens and local governments to consider bio-enzymes as a potential option, systematic research to integrate bio-enzyme-based approach into the existing municipal waste framework is scanty. Usage of bio-enzymes represent a clear shift from linear, top-down models of waste management toward restorative, circular, low-carbon-based waste management system [32].

3 Research Methodology

This chapter presents qualitative case study of three green brands—Eco365, Jalodbust, bluNgreen. The data sources were examined to comprehend the entrepreneurial efforts undertaken by these brands which contribute substantially to the Indian urban environmental solutions.

Rationale Behind the Selection of the Brands

The three green brands, through their eco-focus and sustainable social impact, emerged as potential subjects for the qualitative study.

Eco365: Ideated by Mr Surana in 2010, Eco365, through its retrofitted taps, urinals and the showerheads, has saved 3,500 crore liters of water which is equivalent to INR 150 as per the IndianStartupTimes [33]. This indicates measurable outcomes as far as the level of saved water is concerned which makes Eco365 a potential brand to be studied under urban sustainability [34].

Jalodbust: Engineering for Change (E4C) a non-profit organization that empowers the global engineering workforce to improve the quality of life in underserved communities published on its website about Jalodbust 's products which were aligned with SDG 6 [35]. Jalodbust's unique emphasis on technology-driven and decentralized urban sanitation solution which integrates the public health and environmental engineering makes it an ideal brand to be a part of the descriptive case study-based chapter [36].

blu N green: The BioBliss Series products from bioenzyme-based blu N green are made by India's first bio-enzyme manufacturing company to possess ISO & GMP, Green & Organic ZED certification which indicate world-class parameters of quality, eco-sustainability and Zero Defect Zero Effect respectively. As blu N green is an emerging green business that interfaces bio-enzyme wisdom with cutting-edge science to provide green alternatives to chemical homecare, it becomes a perfect case of sustainable urban entrepreneurship. Though this brand is in its nascent stage, its contribution to sustainability is unquestionable.

The selection may not be exhaustive but environmentally impactful brands are chosen.

The chapter strives to address two major research questions.

1. How do the bio-enzyme based startups use sustainable innovation to address the problem of urban waste management in India?
2. How do the principles of circular economy and sustainable development goal-aligned strategies influence the performance, inclusiveness, and sustainability of the green businesses?

The qualitative analysis is significantly aligned with major theoretical frameworks like The Triple Bottom Line approach which is used to evaluate sustainability performance in the environmental, economic, and social spheres, Rogers' Diffusion of Innovations Theory which provides perspective into the adoption pathways of bio-enzyme solutions and the Effectuation framework which gives an understanding about the entrepreneurial strategies of green startups in politically and economically uncertain conditions.

3.1 Data Collection

Data were compiled from Company webpages with authentic information about the sustainable vision, mission and values of the brands, stories or narratives in the reputable media and on shows, and publicly accessible external directory reports. Data gathered were related to sustainability goals and the reach of these brands in the eco-system they operate in. Although secondary in nature, these data provide credible and reliable insights into the sustainability initiatives of the companies. The publically reviewed information platforms were considered for reliable data. The sustainability claims of the brands in the media articles were cross-verified with the official websites of the brands enabling a context-rich and rigorous analysis about each brand.

3.2 Analytical Framework

This chapter analyses the green intentions expressed by the brands specifically in terms of their alignment with Sustainable Development Goals proposed by the UN. The framework is based on qualitative, interpretive analysis to gain deeper insights about the strategies followed by the brands like environmental responsibility and stakeholder engagement. While the method does not use primary data, it helps contextualize how these brands function in and around the larger green economy (Fig. 1).

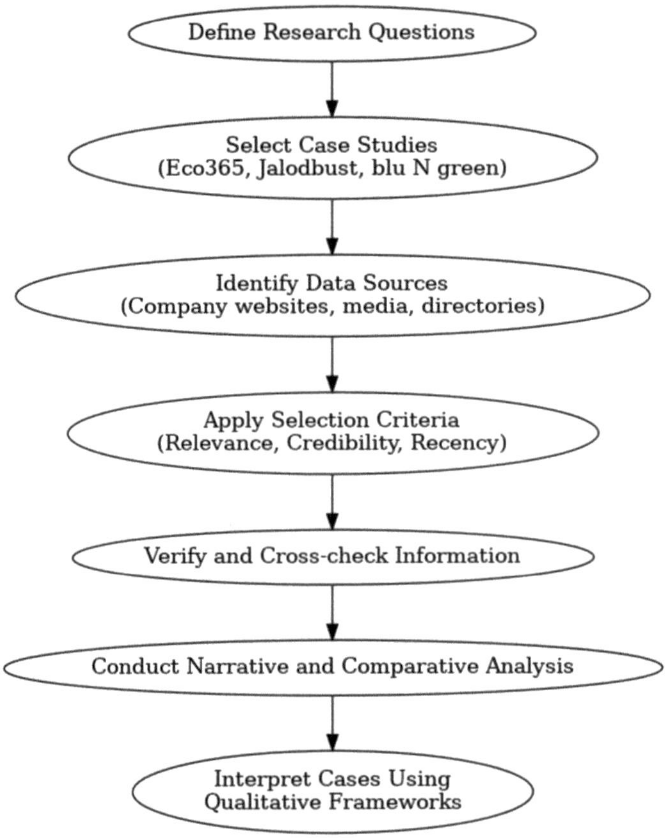

Fig. 1 Stepwise representation of the qualitative methodology adopted for case-based analysis of green startups in India

3.3 Limitation

It is exclusively based on secondary sources such as company reports and trustworthy media reports. Although this method enables a holistic overview of eco-entrepreneurship in India, it is limited in the absence of primary fieldwork or interviews. Impact is thus derived from reported data and stakeholder stories rather than observational evidence. Further research should triangulate these findings through ethnographic or survey-based methods to develop a more communicative model of consumer and organizational behavior.

4 Bio-enzyme Startups and Urban Applications Case Studies

The growing global focus on sustainable urban development has sparked a wave of startups that are harnessing bio-enzymes as nature-friendly solutions to manage urban waste. These bio-enzymes, which come from the anaerobic fermentation of organic materials, provide an eco-conscious alternative to conventional chemical treatments by speeding up waste decomposition and reducing environmental pollution [37]. These innovative startups showcase the blend of ecological sustainability and business potential, marking a fresh approach to sustainable entrepreneurship. In the realm of smart cities, these companies are crucial as they weave cutting-edge biotechnologies into municipal waste management, boosting efficiency and lowering the carbon footprint. Additionally, bio-enzymes align with circular economy principles by facilitating resource recovery and encouraging waste repurposing. This dual emphasis on caring for the environment while seizing economic opportunities resonates with modern policy initiatives, like India's Swachh Bharat Mission and Smart Cities program. As a result, the rise of bio-enzyme startups highlights the power of green innovations in promoting urban sustainability and fostering inclusive economic growth [38].

4.1 Harnessing the Power of Bio-enzymes for a Sustainable Living the Case of blu N Green

A great example of women-led grassroots entrepreneurship making a difference in urban sustainability is the Indian startup blu N green. Apart from formulations, blu N green's home-grown raw material sourcing, zero chemical runoff assurance, and skin-safe promise are entirely in alignment with urban sustainability and circular economy principles. The brand showcases bio-enzymes and the other organic ingredients of the products. This means they are biodegradable, free from harsh chemicals, and planet-safe [39].

Research from urban households, schools, and businesses shows that these bio-enzyme-based cleaners have started to replace traditional chemical cleaners, helping to cut down on environmental pollutants and health risks [40]. The success of blu N green highlights the promise of certified, nature-friendly products in promoting sustainable consumption and emphasizes its importance in fostering green urban transitions. The natural, plant-based ingredients used in blu N green BioBliss Liquid Detergent like rose bio-enzyme, soapnut shell, xanthan gum and coconut-based surfactants are not only biodegradable but also clean effectively and give out waste water that can nourish soil microorganisms.

Urban Impacts of bluNgreen: Traditional cleaning products often contain chemical surfactants that can harm the environment, leading to issues like water pollution, toxicity for aquatic life, and the lingering presence of harmful residues. By choosing bio-enzyme alternatives, these households make a conscious effort to align their buying habits with sustainable waste management principles [41]. This brand strategy emotionally connects with the eco-loving consumers with a strong brand message. Awareness and education play a crucial role in making this transition successful. It is essential to inform consumers about the dangers of chemical-laden cleaning agents, which can emit volatile organic compounds (VOCs), pose respiratory risks, and contribute to long-term environmental harm. Awareness campaigns that use evidence-based communication strategies are key to promoting the benefits of bio-enzymes, such as their biodegradability and non-toxicity.

4.2 Eco365—Institutionalization of Bio-enzyme Adoption

Eco365, a Bengaluru-based startup, caters to various sectors, including hospitality, co-working spaces, and public sanitation systems and boosts sustainability in its own way. Products like Aerator Nozzles and tap flow restrictors from Eco365 can save 75% of the water wastage without compromising on the performance. The aerator covers wide area and minimizes water wastage. This leads in fresh water resource conservation and reduction in household carbon trails. The product range from ECO365 features several water-savers and compostable plastic product categories crafted with sustainability goal [42]. Eco365 makes sustainable living a possibility through its innovative compostable packaging and water saving products which are more than just a fashion label. The product portfolio at Eco365 solves urgent environmental issues like plastics and water shortages. By integrating functionality with eco-friendliness, Eco365 plays its part in conserving resources and helps make a genuine contribution, to both community and planet.

Urban Outcomes of Eco365: One of the standout benefits of bio-enzyme cleaners is their ability to reach into tiny crevices, tackling stains and odors that traditional chemical cleaners often overlook. Additionally, these enzymes keep working even after application, offering ongoing cleaning power, and reducing the need for frequent maintenance [43].

4.3 Jalodbust—A Bio-enzyme-Based Sludge Controller

Jalodbust, nurtured by the Atal Innovation Mission, is revolutionizing the sanitation tech world by effectively handling the pressing issues of fecal sludge management in urban areas that often get overlooked. This startup uses a treatment process that harnesses the power of natural microbes to break down organic waste more quickly, offering a greener and more effective alternative to traditional chemical methods [44]. The process further, catalyzes and biodegrades organic waste, providing a more sustainable and low-impact alternative to traditional chemical cleaners. Jalodbust promotes sustainability through integration of mechanization and decentralization to provide safe and dignified jobs in sanitation, which facilitates community equity and resiliency during health crises such as COVID-19. This integration intersects several Sustainable Development Goals like clean water, sanitation, decent work and economic growth (SDG 6 and SDG 8), reducing inequalities, good health and wellbeing—against urban sanitation's system challenges. With the mobile treatment units, Jalodbust brings sanitation solutions right to the source and helps ease the burden on overloaded sewage systems which prevents environmental pollution and provides flexibility [45].

Urban impacts of Jalodbust: Jalodbust has demonstrated significant progress in addressing urban sanitation challenges through its innovative bio-enzyme-based fecal sludge treatment technology. Empirical evidence indicates that their pre-treatment process effectively reduces pathogen load by up to 40%, contributing substantially to improved public health outcomes in densely populated and underserved communities. The initiative's impact spans more than 30 slums and peri-urban wards, highlighting its scalability and adaptability across diverse socio-economic and infrastructural contexts [46]. Jalodbust's decentralized treatment model exemplifies a practical, nature-based solution that not only mitigates environmental pollution but also empowers disadvantaged communities by addressing their sanitation needs sustainably. This aligns with broader Sustainable Development Goals, particularly SDG 6 on clean water and sanitation, and offers a replicable framework for integrating bio-innovations into urban waste management ecosystems [47].

4.4 Learnings and Common Threads

Across these case studies, two deep patterns emerge:

a. **Context-Specific Scalability**: The bio-enzyme formulations and delivery systems are designed to meet the specific waste types and volumes found in different environments, whether it is a home or a municipal sewer system. This flexibility allows these companies to tackle local sanitation issues effectively [48].
b. **Strategic Eco-Brand Positioning**: Entrepreneurs craft engaging brand stories that are rooted in scientific proof and community benefits. They highlight sustainability certifications along with real-life success stories that connect with eco-conscious consumers and policymakers.

4.5 Tech-Enabled Innovations Boosting the Sustainable Bio-enzyme Brands

While bio-enzymes are at the heart of these startups, their real success comes from brilliantly blending advanced technologies that turn them into all-encompassing, tech-savvy sustainability solutions [49].

blu N green, harnesses digital platforms to monitor product lifecycles, streamline order fulfillment, and connect with consumers through awareness campaigns on social media and e-commerce sites. Blu N green also uses advanced bio-enzyme technology, using biologically fermented microbial cultures to break down soil, stain and odor naturally.

Jalodbust stands out in the sanitation-tech arena by using mobile bio-enzyme treatment units that come with real-time monitoring tools, to ensure efficient fecal sludge management in underserved urban areas. Jalodbust's vacuum unit is portable, battery-operated and supported by hydraulic shockwaves to remove dense human waste from septic tanks and manholes that allows no human touch [50]. It integrates smart control panel IoT dashboards and lithium-ion batteries with its product SaniPreneur to safely operate in the urban areas.

Eco365 installs IoT-enabled, waterless urinals using bio-enzyme with the back-gas barrier technology and saves 100% water per use. It adopts aerators pressure compensating and flow regulators that can limit water consumption at taps and showers without limiting user experience.

By pairing nature-based products with a strong technological backbone, these companies not only tackle urban waste and sanitation issues but also set up scalable, replicable, and resilient models for sustainable urban entrepreneurship in our digital world [51].

5 Role of Bio-innovation in Cities—Assessing Gaps in Policy and Ecosystem Enablers

This section of the chapter evaluates the facilitating ecosystem for the eco-based brands taking insights from the 3 chosen cases, focusing on the policy framework and operational deficiencies. It further evaluates how the existing certifications and the capacity-building process integrate with the ground-level innovation efforts and where they prove inadequate.

The widespread adoption of bio-enzymes is challenged by structural constraints like fragmented municipal waste systems, inconsistent regulatory landscape, and low public interest. Fixing decentralized, environmentally sustainable solutions into the rigid urban sanitation mechanism will require established system of waste collection, segregation and processing, and will therefore be beyond the level which an innovation alone can handle. While partnerships with municipal corporations and

CSR programs are an ideal approach, they prove ineffective because of the bureaucratic inertia, fragmented funding channels and under-trained sanitation workforce. Bio-scientific solutions like bio-enzyme treatments need to be integrated into urban waste management systems in India with a long-term, systemic approach [52]. The potential of bio-enzymes to change city-level waste systems is tremendous, but the road ahead is both filled with structural enablers and systemic barriers [53].

Programs such as Startup India and Atal Innovation Mission (AIM) have fostered sustainability-focused entrepreneurship. AIM's incubators and Atal Tinkering Labs are leading the way for eco-technological innovation right from the grassroots to the youth [54]. These tools enable life-science entrepreneurs to prototype bio-enzyme alternatives to chemical-intensive cleaning and treatment processes. Similarly, the Smart Cities Mission, launched with the vision of leveraging technology as an enabler for sustainable urbanization, now serves as a powerful policy anchor for integrating decentralized, nature-based solutions such as bio-enzymes into urban planning, particularly in sanitation and 'waste-to-resource' models [55].

Indian cities like Indore and Pune follow the principles of decentralized composting and green cleaning and adopt them in their urban governance systems through vendor certifications and enabling byelaws while many other urban local bodies still follow the chemically-intense processing models of waste management. This disparity among the different locations and regions greatly prevents the process of bio-enzymes adoption in the larger urban framework thereby leading to the lack of awareness among the mainstream public.

Certifications and assessments like ISO 14001 for environmental management system, GreenPro for eco-labeling, Life Cycle Assessment to assess environmental impact in a product's lifespan, and Environmental Impact Assessments to evaluate the environmental risk, enable bio-enzyme startups to access green markets, but integration with urban policy and governance are essential.

For the bio-enzyme innovations to bear full potential in India, a roadmap embedding urban sanitation policy and certification regimes besides local urban governance is required [5]. This would include:

- Prescribing Green Procurement to all the Urban Local Bodies (ULBs) to drives demand for sustainable, eco-friendly products
- Streamlined certification of bio-enzyme-based products to ease green market access and build trust among the users, a model followed by the government of Telangana, a South Indian state, which encourages eco-startups through the T-Hub with tools to navigate certifications
- Providing GST rebates and subsidies for businesses that are green conscious to reduce cost and motivate green entrepreneurship
- Mainstream bio-enzyme products to embed their usage in urban sanitation guidelines, including Swachh Bharat 2.0.

These reforms are not mere technical compliance, but move beyond that to prove as critical enablers for mainstreaming climate-resilient and inclusive urban planning.

Through concerted policy support, bio-enzymes can develop from small, sporadic innovations into potent tools for public health enhancement, urban environmental redemption and green enterprise creation, taking India a few steps closer to SDG 11 (Sustainable Cities) and SDG 12 (Sustainable Consumption and Production). With institutional commitment, regulatory control, consistent campaigns to shift public behavior toward cleaner solutions, policy frameworks, and the stakeholders' involvement, measurable success can be registered, but over a considerable period of time.

6 Strategic Recommendations

Policy Integration for Systemic Adoption: The first step in making bio-enzymes a part of everyday life is getting them officially recognized in both national and local urban development policies. Initiatives like the Smart Cities Mission and Swachh Bharat Abhiyan should clearly incorporate bio-enzyme-based sanitation and cleaning innovations into their environmental sustainability goals.

Establishing Robust Standards and Certifications: To gain credibility and build consumer trust, a standardized framework for certifying bio-enzymes is required. Government agencies like BIS (Bureau of Indian Standards), FSSAI (Food Safety Standards Authority of India), and third-party certifying organizations should work together to simplify eco-labeling, organic verification, and ISO certification. This standardization will make it easier for bio-enzyme-based products to enter public procurement systems and help cut down on greenwashing. A clear labeling process will empower consumers to make informed, sustainable choices.

Strengthening Workforce Capacities: Urban sanitation workers, municipal engineers, and facility managers need specialized training programs that cover the science, application, and benefits of bio-enzymes.

Promoting Public Awareness and Changing Behaviors: Mass media campaigns that highlight real-life success stories, present environmental data, and showcase the cost benefits can help change how people view bio-enzyme based products, moving them from a niche market to a more mainstream option. Social media influencers, resident welfare associations (RWAs), and schools can act as effective channels for spreading awareness.

The road ahead: Bio-enzymes startups are spearheading a silent revolution in waste management in cities, where natural, microbial products from processing organic waste are replacing chemical-laden products. National policies such as Startup India, Swachh Bharat 2.0, and Smart Cities Mission must actively endorse enzyme-based sanitation systems. Startups in sustainability realm, need support in terms of sector-specific incubation, streamlined access to certification like ISO 14001 and GreenPro and inclusion in green public procurement. Municipal bodies must train their staff on the usage eco-friendly solutions to facilitate a deeper understanding and conviction about their planet-friendly features.

The theoretical perspectives through the Triple Bottom Line Approach, Roger's diffusion of Innovation and the entrepreneurial theory, together signify the long-time process of systemic change in urban sustainability, where innovations requires support in terms of suitable ecosystems, behavioral shifts, and policy framework.

Circular economy principles and Sustainable Development Goals (SDGs) are not just theoretical frameworks, but are increasingly becoming the operational blueprints for green business models. Bio-enzyme companies and brands deserve economic and social support as they turn waste to wealth, recycle natural resources and minimize ecological damage [56].

The bio-enzyme based companies can further drive this alignment by leveraging cutting-edge technologies to scale natural cleaning in the urban built environments. By facilitating customizable and more efficient degradation, enzyme engineering and microbial optimization enable high-purity, high-potency products with improved specificity. Automated application at public and institutional settings using smart dispensing and IoT dosing systems enable precision and minimum wastage. Innovations in biodegradable packaging like cornstarch or seaweed films help zero-waste delivery and circular use. By enhancing penetration on urban surfaces, nano-biotech surface formulations ensure deep and continuous cleaning. Digital tracking platforms and AI dashboards allow impact monitoring and awareness, which help both consumers and institutions to measure ecological benefits. Greywater integration and septic pre-treatment enable water reclamation opportunities and expand cleaning into the water management systems to support hygiene. By utilizing cold-chain stabilization and microencapsulation for enzymes, we enable climate-resilient storage of the enzymes along with assured shelf-life as well as activity at a wide range of conditions. Taken together, these innovations also enable bio-enzyme brands to unleash the potential for sustainable cities. With the right support, bio-enzymes can transcend the present status as upcoming innovation and be considered as a transformative tool and a catalyst for a resilient, inclusive India

References

1. World Bank (n.d.) Solid waste management. https://www.worldbank.org/en/topic/urbandevelopment/brief/solid-waste-management
2. Kaza S, Yao L, Bhada-Tata P, Van Woerden F (2018) What a waste 2.0: a global snapshot of solid waste management to 2050. World Bank Group
3. Patel J, Mujumdar S, Srivastava VK (2023) Municipal solid waste management in India—current status, management practices, models, impacts, limitations, and challenges in future. Adv Environ Res 12(2):95–111. https://doi.org/10.12989/aer.2023.12.2.095
4. de Lemos FP, Filho HSC, Foresti E (2022) Enzymatic hydrolysis of waste fats, oils and greases (FOGs): status, prospective, and process intensification alternatives. Chem Eng Processing—Process Intensif 175:108930. https://doi.org/10.1016/j.cep.2022.108930
5. Kumar R, Sharma A, Gupta V (2021) Enzyme-based solutions for sustainable waste treatment: emerging trends and applications. J Environ Manag 287:112275. https://doi.org/10.1016/j.jenvman.2021.112275

6. Paul D, Malik S, Mishra DK, Teotia A (2023) A study on effectiveness of ecolabels in the fast moving consumer goods sector. IOP Conf Ser: Earth Environ Sci 1161(1):012004. https://doi.org/10.1088/1755-1315/1161/1/012004
7. Chakraborty D, Chatterjee S, Althuri A, Palani SG, Venkata Mohan S (2023) Sustainable enzymatic treatment of organic waste in a framework of circular economy. Biores Technol 379:128487. https://doi.org/10.1016/j.biortech.2022.128487
8. Mukherjee M, Zwarteveen M (2020) Women entrepreneurs and environmental sustainability: empowerment through green business initiatives in urban India. World Dev 128:104841. https://doi.org/10.1016/j.worlddev.2019.104841
9. United Nations (2015) Transforming our world: the 2030 agenda for sustainable development. United Nations. https://sdgs.un.org/2030agenda
10. Singh G, Kumar R, Tiwari B, Kumar M (2024) Exploring garbage enzymes as novel biocatalyst for enhancing bioprocess performance in composting. J Clean Prod. https://doi.org/10.1016/j.jclepro.2024.100591
11. Elkington J (1997) Cannibals with forks: the triple bottom line of 21st century business. Capstone Publishing
12. Rogers EM (2003) Diffusion of innovations, 5th ed. Free Press
13. Sarasvathy SD (2001) Causation and effectuation: toward a theoretical shift from economic inevitability to entrepreneurial contingency. Acad Manag Rev 26(2):243–263. https://doi.org/10.5465/amr.2001.4378020
14. blu N green (n.d.) About us. blu N green. https://blungreen.in/pages/about-us
15. Saluja S, Gaur A, Somani P, Sharma A (2024) Innovative approach to waste management: utilizing stabilized municipal solid waste in road infrastructure. Environ Sci Pollut Res 31:10346–10358. https://doi.org/10.1007/s11356-024-10346-10358
16. Gulhane PA, Wankhade AV (2019) Fruit and vegetable waste bioenzyme: a novel, low-cost eco-friendly cleaner. Environ Technol Innov 16:100466. https://doi.org/10.1016/j.eti.2019.100466
17. Moretto G, Russo I, Bolzonella D, Pavan P, Majone M, Valentino F (2023) Anaerobic fermentation of organic solid waste: recent updates in substrates, products, and the process with multiple products co-production. Environ Res 233:116444. https://doi.org/10.1016/j.envres.2023.116444
18. Scheibel DM, Gitsov IPI, Gitsov I (2024) Enzymes in "green" synthetic chemistry: laccase and lipase. Molecules 29(5):989. https://doi.org/10.3390/molecules29050989
19. Prakash G, Singh PK, Ahmad A, Kumar G (2023) Trust, convenience and environmental concern in consumer purchase intention for organic food. Span J Mark—ESIC 27(3):367–388. https://doi.org/10.1108/SJME-09-2022-0201
20. Szpilko D, de la Torre Gallegos A, Jimenez Naharro F, Rzepka A, Remiszewska A (2023) Waste management in the smart city: current practices and future directions. Resources 12(10):115. https://doi.org/10.3390/resources12100115
21. Yadav H, Kumar S, Sharma A (2024) Integrated microbial–enzyme systems for sustainable urban organic waste management: Scalability and infrastructure adaptability. Biores Technol 397:128922. https://doi.org/10.1016/j.biortech.2024.128922
22. Sharma P, Verma A (2023) Sustainable innovation pathways for urban sanitation in India: opportunities under Swachh Bharat and AMRUT missions. J Urban Manag 12(1):45–58. https://doi.org/10.1016/j.jum.2023.02.004
23. Yan F, Xu X, Li H et al (2023) Beyond composting basics: strategic guidance for IoT integration in urban composting ecosystems. Sustainability 16(23):10332. https://doi.org/10.3390/su162310332
24. Hati S, Patel A, Mandal S (2020) Bioenzymes from fruit waste: characteristics and environmental applications. Bioresour Technol Rep 11:100524. https://doi.org/10.1016/j.biteb.2020.100524
25. Zhao L, Chen F, Liu Y (2023) From waste to wealth: Innovations in organic solid waste composting. Environ Res 229:115977. https://doi.org/10.1016/j.envres.2023.115977
26. Santoro G, Vrontis D, Thrassou A, Dezi L (2018) Sustainable innovation: a systematic review. J Clean Prod 179:704–724. https://doi.org/10.1016/j.jclepro.2017.11.029

27. Wang H, Zhang Y, Li J (2024) Conversion of organic solid waste into energy and functional materials using biochar catalyst: bibliometric analysis, research progress, and directions. Appl Catal B 340:123223. https://doi.org/10.1016/j.apcatb.2023.123223
28. Kim J, Park S (2023) Regulatory and market barriers for microbial enzyme cleaners: a case study of the EU and US. J Clean Prod 45(3):115–129. https://doi.org/10.1016/j.jclepro.2023.115129
29. Maalouf A, Mavropoulos A (2023) Re-assessing global municipal solid waste generation: projections and management implications. Waste Manag Res 41(4):212–225. https://doi.org/10.1177/0734242X221074116
30. Ray P (2024) Exploring the role of smart city strategies in fostering entrepreneurship in small and medium-sized cities of India: a case study of Indore and Satna. ResearchGate. https://doi.org/10.1080/13673882.2023.00001015
31. Chakraborty D, Chatterjee S, Althuri A, Palani SG, Venkata Mohan S (2023) Integrated bio-enzymatic treatment of domestic greywater: efficiency, safety, and potential for decentralized reuse. Biores Technol 390:128651. https://doi.org/10.1016/j.biortech.2023.128651
32. Cheela VRS, John M, Biswas WK, Dubey B (2021) Environmental impact evaluation of current municipal solid waste treatments in India using life cycle assessment. Energies 14(11):3133. https://doi.org/10.3390/en14113133
33. https://www.indianstartuptimes.com/interviews/an-insightful-conversation-with-goutam-surana-founder-eco365-championing-water-conservation-and-sustainability-in-india
34. ECO365 (n.d.) Eco365 sustainable products. Prakati. https://prakati.com/d/eco365/
35. Engineering for Change (n.d.) Jalodbust sludge remover. https://www.engineeringforchange.org/solutions/product/jalodbust-sludge-remover/
36. Jalodbust (n.d.) About me. https://jalodbust.com/about-me/
37. Vilar DS, Bilal M, Bharagava RN et al (2022) Lignin-modifying enzymes: a green and environmentally responsive technology for organic compound degradation. J Chem Technol Biotechnol 97(2):327–342. https://doi.org/10.1002/jctb.7114
38. Arun C, Sivashanmugam P (2015) Investigation of biocatalytic potential of garbage enzyme and its influence on stabilization of industrial waste activated sludge. Process Saf Environ Prot 94:471–478. https://doi.org/10.1016/j.psep.2014.10.008
39. blu N green (2024) Taking the eco-friendly path forward with natural alternatives [Status update]. LinkedIn. https://www.linkedin.com/feed/update/urn:li:activity:7348584016775737345/
40. Asadi V, Oh S-G, Kim H-J, Kim D-H (2023) Effectiveness of environmentally friendly enzymatic cleaning solutions in reducing chemical pollutant exposure in urban households and schools. J Clean Prod 385:135895. https://doi.org/10.1016/j.jclepro.2023.135895
41. blu N green (2024) #SustainableLiving is not a trend, it's a choice. Let's choose the Blu way [Status update]. LinkedIn. https://www.linkedin.com/posts/blungreen_blungreen-sustainableliving-ecolovers-activity-7330467654257758209-pNj-/
42. ECO365 (n.d.) Company profile—ECO365 private limited. The company check. https://www.thecompanycheck.com/company/eco-365-private-limited/U29190KA2020PTC131783
43. ECO365 (n.d.) About compostable plastic. https://eco365.co.in/compostable-plastic/
44. Jalodbust (n.d.) Agricultural benefit on environment. https://jalodbust.com/agriculture-benefit-on-environment/
45. YourStory (2020) This startup has mechanised manual scavenging to help 2 lakh people live with dignity. https://yourstory.com/socialstory/2020/11/jalodbust-mechanised-manual-scavenging-20-lakh-people
46. Jalodbust (n.d.) Impact on humanity. https://jalodbust.com/impact-on-humanity/
47. https://www.outlookindia.com/national/clean-up-sanitation-ecosystem-magazine-331536
48. Xu Q, Li J, Chen X (2023) Current progress, opportunities and challenges of developing green disinfectants for the remediation of disinfectant emerging contaminants. Environ Sci Ecotechnology 12:100371. https://doi.org/10.1016/j.esg.2023.100371
49. Leong HY, Chang CK, Khoo KS et al (2021) Waste biorefinery towards a sustainable circular bioeconomy: a solution to global issues. Biotechnol Biofuels Bioprod 14:87. https://doi.org/10.1186/s13068-021-01939-5

50. https://yourstory.com/socialstory/2020/11/jalodbust-mechanised-manual-scavenging-20-lakh-people
51. Begum YA, Kumari S, Garg MC (2024) A review on waste biomass to energy: integrated thermochemical and biochemical conversion for resource recovery. Environ Sci: Adv 3:1197–1216. https://doi.org/10.1039/D4VA00109E
52. Chouhan PK, Bahuguna A (2024) Cleaning Bharat (India): a scoping review of perceptions regarding how Swachh Bharat can continue to improve hygiene across the nation. Int J Environ Res Public Health 21(5):17. https://doi.org/10.3390/ijerph21010017
53. Surana K, Singh A, Sagar AD (2020) Strengthening science, technology, and innovation-based incubators to achieve sustainable development goals: lessons from India. https://arxiv.org/abs/2005.13138
54. Pansera M, Owen R (2018) Framing inclusive innovation within the discourse of development: insights from case studies in India. Res Policy 47(1):23–34. https://doi.org/10.1016/j.respol.2017.09.007
55. Shruti S, Singh PK, Ohri A (2021) Evaluating the environmental sustainability of smart cities in India: the design and application of the Indian smart city environmental sustainability index. Sustainability 13(1):327. https://doi.org/10.3390/su13010327
56. Yadav P, Sinha B (2021) Turning waste into wealth: sustainable urban waste management through green entrepreneurship in India. J Clean Prod 278:123805. https://doi.org/10.1016/j.jclepro.2020.123805

Urban Commons Reimagined: Revitalizing Wet Markets Through Digital Culture and Generational Exchange

MinJoo Baek and Iasef Md Rian

Abstract Traditional wet markets in East Asia are experiencing accelerated decline, driven by the growth of e-commerce, top-down urban redevelopment, and widening intergenerational disconnection. Yet, these markets remain critical socio-cultural infrastructures, supporting local economies, ecological resilience, and intangible heritage. This chapter reimagines wet markets as urban commons, proposing that digital tools, when grounded in community narratives and co-creation, can mediate between tradition and innovation. Drawing on comparative case studies from Suzhou, Hangzhou, Singapore, Jeonju, London, and other global cities, the chapter examines how QR storytelling, mobile apps, blockchain traceability, and social media platforms can enable inclusive cultural participation and youth engagement. By integrating digital placemaking, participatory urbanism, and commons theory, this work offers a transferable framework for revitalizing wet markets as resilient, intergenerational, and culturally embedded public spaces.

Keywords Wet markets · Digital placemaking · Urban commons · Youth co-design · Cultural sustainability · Participatory heritage

1 Introduction

1.1 Background

Wet markets in East Asia stand at a critical juncture. Once integral to the everyday rhythms of urban life, providing affordable produce, supporting informal economies, and sustaining intergenerational social interaction, they are now facing rapid decline.

M. Baek (✉) · I. Md Rian
Department of Architecture, Xi'an Jiaotong-Liverpool University, Suzhou, China
e-mail: minjoo.baek@xjtlu.edu.cn

I. Md Rian
e-mail: iasef.rian@xjtlu.edu.cn

V. Chand Sharma et al. (eds.), *Tech-Enabled Urbanism and Entrepreneurship for Inclusive Cities*, Lecture Notes in Civil Engineering 779,
https://doi.org/10.1007/978-981-95-3873-7_9

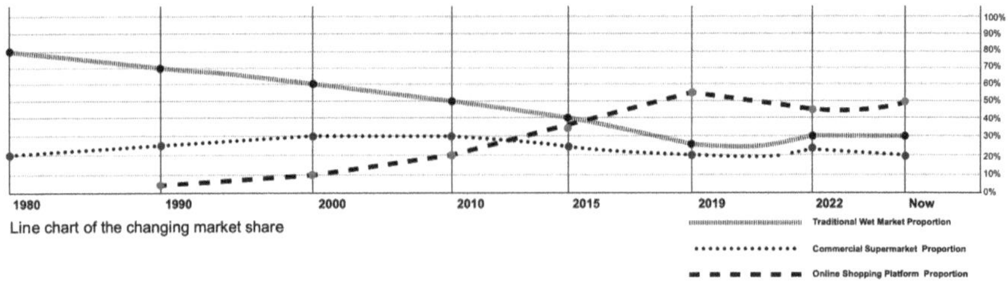

Fig. 1 Decline in wet market participation in China (2010–2024). *Source* Compiled from China urban market report (2023) and digital market transition survey (2024)

The pressures of e-commerce, urban redevelopment, shifting consumer expectations, and an expanding generational divide have accelerated their marginalization [1, 2]. Younger, digitally native consumers increasingly favor online grocery platforms, drawn by features such as mobile payment, real-time updates, and traceable sourcing. Consequently, traditional markets are often perceived as unhygienic, inconvenient, and digitally obsolete [3, 4]. Figure 1 visualizes data from municipal surveys showing a 20% drop in daily footfall in traditional markets between 2010 and 2024. The figure contextualizes the urgency of revitalization strategies.

Yet, wet markets remain vital components of urban food systems and cultural infrastructure. They offer not only locally sourced, minimally processed produce, but also spaces of everyday sociability, place attachment, and intangible cultural heritage [5–7]. Their decline signals not merely a shift in consumer behavior but a deeper erosion of community-based urban life.

This chapter contends that the revitalization of wet markets lies not in their wholesale modernization but in their reimagining as urban commons, shared civic resources grounded in cultural memory, participatory governance, and intergenerational stewardship [8]. Through the lens of digital placemaking, the chapter explores how technologies such as QR storytelling, social media, mobile apps, and augmented reality can mediate the evolving relationships between youth, place, and heritage [9, 10]. Far from replacing traditional markets, these tools offer opportunities for participatory placemaking, youth co-design, and community-led innovation.

Drawing on comparative case studies in Suzhou, Hangzhou, Singapore, Jeonju, and London, this chapter critically investigates how digital culture can help reposition wet markets as hybrid, resilient, and inclusive urban commons, sites of both cultural continuity and civic transformation.

1.2 Current State of the Research

Over the past decade, research has increasingly recognized the enduring significance of wet markets in supporting sustainable diets, local economies, and cultural resilience across Asia and beyond [6, 7, 11, 12]. Wet markets remain primary sources of fresh, affordable food for both urban and rural communities, playing a pivotal role in shaping food environments and supporting community well-being [13]. Despite these contributions, wet markets face mounting challenges, including spatial marginalization, competition from supermarkets and e-commerce, and waning engagement among younger generations [14–16].

Recent public health concerns, particularly in the wake of the COVID-19 pandemic, have intensified scrutiny of hygiene and biosecurity practices in these markets, prompting calls for modernization and regulatory reform [17]. While such interventions can improve cleanliness and safety, they also risk eroding traditional practices and social bonds that underpin the resilience of wet markets [18].

Current research highlights the need for participatory approaches that engage diverse stakeholders in the revitalization of wet markets. Community-driven innovations and targeted policies can help balance the preservation of cultural authenticity with the demands of modernization [19]. However, gaps remain in understanding how digital technologies and participatory strategies can be leveraged to foster inclusivity and intergenerational engagement. Addressing these gaps, this chapter examines how wet markets might be reimagined as inclusive urban commons, balancing innovation with the preservation of cultural continuity.

1.3 Research Gap and Questions

While a growing body of research addresses wet market modernization, food safety, and the rise of digital urbanism, most studies treat these themes separately, focusing either on technological upgrades and regulatory reforms or on the socio-cultural dimensions of market decline [20]. There is a notable lack of integrated frameworks that examine how digital technologies can be strategically employed to sustain cultural continuity and foster civic participation in traditional markets. In particular, the intersection of digital placemaking and participatory heritage in wet markets remains underexplored, with limited empirical research on how digital storytelling, co-creation, and youth engagement can bridge generational divides and support inclusive urban commons [21–23].

A significant gap concerns the role of youth as active co-designers and custodians of urban memory. Although young people increasingly engage with heritage through digital and informal means, their agency in shaping digital placemaking and heritage revitalization is rarely foregrounded in academic or policy discourse. Recent research highlights that while digital platforms are being used to engage younger generations with cultural heritage, youth involvement often remains limited to participation rather than genuine co-design or custodianship. Scholars call for more innovative, youth-driven approaches to heritage engagement and urban memory to ensure intergenerational continuity and inclusivity [24–26]. Furthermore, many existing studies rely on short-term or localized case analyses, limiting the generalizability and long-term understanding of digital interventions' social, cultural, and economic impacts. There is also insufficient attention to the risks of digital exclusion, gentrification, and fragmented regulatory frameworks that may undermine the inclusivity of market revitalization efforts [20].

So, the research question is: How can digital technologies and participatory design strategies revitalize traditional wet markets by engaging younger generations while preserving cultural authenticity and continuity?

1.4 Aim and Objectives

The primary aim of this research is to develop a conceptual and empirically informed framework for revitalizing traditional wet markets as inclusive urban commons through the use of digital cultural tools and intergenerational design.

a. to achieve this aim, the study pursues five key objectives:
b. to map the challenges and socio-cultural roles of wet markets in East Asia;
c. to analyze youth-led and digitally mediated design interventions that support market revitalization;
d. to compare case studies from Suzhou, Hangzhou (China), Singapore, Jeonju (Korea), London (UK), and other global cities to identify transferable practices;
e. to develop a theoretical model that integrates digital placemaking and commons governance;
f. to recommend participatory strategies for wet market vendors, policymakers, and urban designers.

Through these objectives, the chapter seeks to demonstrate how digital and participatory strategies can sustain the relevance, inclusivity, and cultural vibrancy of wet markets in rapidly changing urban environments.

2 Methodology and Context Selection

This study adopts a comparative, multi-sited case study approach, focusing on wet markets in Suzhou, Hangzhou (China), Singapore, Jeonju (South Korea), London (UK), and additional global cities. These locations were selected to represent diverse trajectories of urban development, digital maturity, and heritage policy, enabling robust cross-contextual analysis [8, 19]. The research design integrates several complementary methods. Case study analysis is used to document and evaluate design-led interventions, such as QR code trails, mobile applications, and youth-led workshops, which have been central to digital and participatory revitalization efforts [27, 28]. Literature synthesis reviews key research literature in urban studies, food geography, heritage theory, and platform urbanism to situate the findings within broader academic debates [9, 29]. Finally, comparative synthesis is employed to identify cross-cutting themes and transferable practices that can inform future revitalization initiatives.

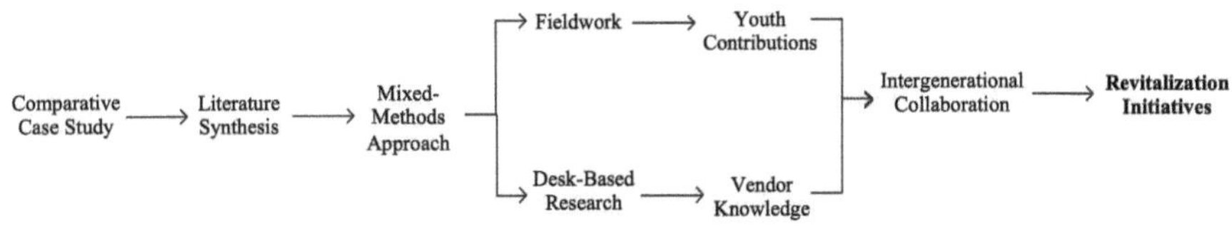

Fig. 2 Mixed-method workflow

This mixed-methods approach combines fieldwork with desk-based research, emphasizing both direct observation and critical analysis of existing literature. Figure 2 specifies the study's sequential mixed-methods design, integrating field observations (Oct 2023 to May 2025) and comparative analysis across five cities. This workflow ensured the inclusion of both embedded market knowledge and generational digital perspectives. Special attention is given to the creative contributions of youth and the embedded knowledge of long-standing market vendors and stakeholders [30]. By foregrounding the roles of both younger participants and experienced actors, the methodology highlights how intergenerational collaboration and local expertise are essential for shaping the future of wet markets as vibrant, inclusive urban commons [28].

Additional cities such as Barcelona, Guangzhou, and Manila were selected based on three criteria: (1) integration of digital technology in traditional markets; (2) recognition in policy or academic discourse as heritage markets; and (3) availability of comparative public data and reports.

Fieldwork was conducted from 2022 to 2024 across four sites, involving 20 + direct observations, informal interviews with vendors, and youth participants. Observational protocols focused on spatial layout, technology integration, vendor-customer interaction, and youth participation.

3 Case Studies

Recent years have seen traditional wet markets in cities such as Suzhou, Hangzhou, Singapore, Jeonju, and London transformed by the integration of digital technologies and participatory design. Interventions such as QR code trails, mobile applications, social media engagement, and digital mapping have enhanced transparency, visitor experience, and cultural storytelling, while also supporting the preservation of local heritage and fostering intergenerational participation [27, 31, 32]. The following case studies illustrate how digital placemaking and creative engagement are revitalizing wet markets as vibrant, inclusive urban commons.

3.1 Suzhou Shuangta Market, China: Digital Tools and Visitor Impact

Suzhou's Shuangta Market exemplifies the successful integration of heritage preservation with digital innovation and creative urban renewal. Recent renovations have focused on blending traditional cultural elements with modern amenities, establishing the market as both a vibrant community hub and a popular tourist destination [32, 33]. A key component of this revitalization has been the implementation of advanced digital infrastructure, including food traceability systems powered by QR codes and digital signage. These tools have enhanced transparency and operational efficiency, providing visitors with accessible, real-time information about product origins, pricing, and vendor details [34]. As shown in Fig. 3, these digital interfaces visibly communicate key product data, making market transactions more transparent and fostering consumer trust.

In addition to digital upgrades, the market's transformation features themed decorations, specialty souvenir shops, cafés, and a bookstore, collectively creating an inviting and dynamic atmosphere that appeals to both residents and tourists [32]. These enhancements have had a measurable impact on visitor numbers: following the upgrades, Shuangta Market reportedly attracts approximately 8,000 visitors on weekdays and up to 12,000 on weekends, marking a significant increase in foot traffic and popularity [32, 33]. Notably, all original vendors have remained at the venue after the renovation, benefiting from the influx of new customers and the expanded community reach [32, 35].

Overall, the Shuangta Market case demonstrates how the thoughtful application of digital tools and creative design can drive urban market revitalization, increase visitor engagement, and sustain the cultural vibrancy of historic districts in contemporary Chinese cities. The success of Shuangta Market serves as a reference point for similar improvement projects in other urban markets, highlighting the potential of digital-cultural interventions to support both economic vitality and heritage preservation [32, 33, 35].

Fig. 3 Product origin and pricing information displayed on digital panels at Suzhou's Shuangta market. *Source* CCTV News 央视关注苏州:走进双塔市集 感受姑苏古城"老味道 [CCTV focuses on Suzhou: Enter the Shuangta Market to experience the "old flavor" of the ancient city of Gusu] (May 25, 2025)

3.2 *Hangzhou Wener Market, China: Storytelling and Digital Transparency*

Hangzhou's Wener Market has emerged as a leading example of how digital technology can be harnessed to enhance transparency and create engaging narratives in urban food markets. The market's recent transformation centers on the integration of interactive digital installations, such as touchscreens at vendor booths, which provide visitors with detailed information about product origins, vendor backgrounds, and cooking techniques. These features are designed to foster trust and enrich the educational experience for market-goers, particularly appealing to younger, tech-savvy consumers who value transparency and authenticity in food sourcing [36, 37]. Figure 4 captures one such installation, where QR-enabled touchscreens invite users to explore vendor stories and culinary heritage in an interactive format.

A notable aspect of Wener Market's digital strategy is the widespread use of QR codes throughout the market (Fig. 4). These codes link visitors to video interviews with vendors, traditional recipes, and additional multimedia content, allowing for a deeper connection between customers and the stories behind the products they purchase [37, 38]. The market also offers a dedicated mobile app that provides real-time updates on promotions, stall locations, and special events, further enhancing convenience and engagement [37].

The combination of digital storytelling and transparency has been recognized as a key factor in revitalizing Wener Market, fostering stronger relationships between vendors and customers, and positioning the market as a model for innovative urban food retail. These digital features have contributed to increased visitor engagement, improved market reputation, and a more vibrant community atmosphere.

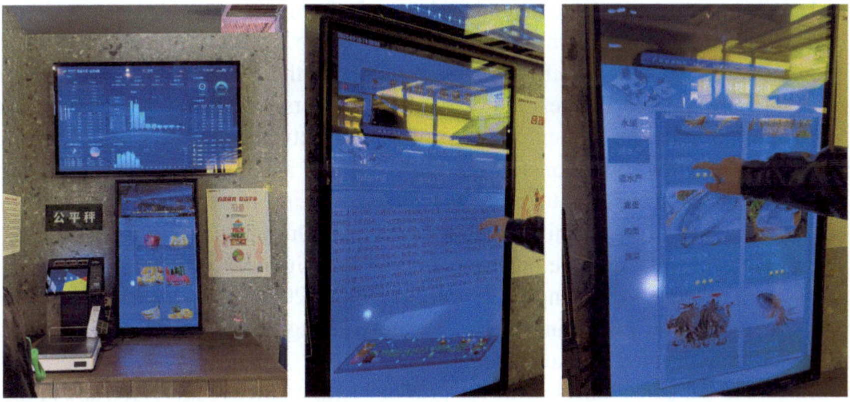

Fig. 4 Interactive QR and touchscreen installation in Wener market. *Source* Author

Fig. 5 Google deployed a mobile operator with a backpack-mounted camera system to collect 360-degree images throughout Singapore's hawker centers. *Note.* Adapted from "Google maps AR launched by Google and Singapore tourism," 2024. *Source* Immersive-Technology[1]

3.3 *Singapore: Digital Marketing, Community Engagement, and AR Tourism*

Singapore's approach to revitalizing its wet markets and hawker centers demonstrates the transformative power of digital marketing, community engagement, and immersive digital mapping technologies. These strategies have not only modernized traditional food retail but also enhanced the cultural and tourism value of these spaces.

Singaporean wet markets, exemplified by Quan Shui Wet Market, have embraced digital marketing to reach new audiences and adapt to changing consumer behaviors. Quan Shui Wet Market, a family-run business, pivoted to online sales and digital engagement in response to declining foot traffic and the challenges posed by the COVID-19 pandemic. By leveraging social media platforms such as Facebook and implementing a dedicated website, the business expanded its reach and improved customer convenience [27]. As shown in Fig. 5, part of this effort included Google's deployment of mobile operators equipped with backpack-mounted cameras to capture 360-degree imagery across hawker centers, contributing to immersive digital content creation and enhancing online visibility. Participation in Enterprise Singapore's Digital Practitioner Programme (DPP), a collaboration with Google (Fig. 5), enabled the business to refine its digital marketing strategy, resulting in a 48% increase in website visits within five months [27, 39]. Online sales now contribute 40–50% of total revenue, and the business serves approximately 500 households daily, with overall revenue increasing by 25% from 2021 to 2024 [27]. This case illustrates how targeted digital marketing and community engagement can drive growth and resilience in traditional market businesses.

Singapore has also pioneered the integration of augmented reality (AR) and digital mapping to enhance the visitor experience at food centers and markets. The Singapore Tourism Board (STB), in partnership with Google, has developed location-based

[1] https://immersive-technology.com/augmentedreality/google-maps-ar-launched-by-google-and-singapore-tourism/.

AR experiences accessible via Google Maps and the Visit Singapore Travel Guide app. These AR features provide interactive tours of iconic food centers such as Maxwell Food Centre, offering visitors immersive storytelling, historical context, and recommendations for must-try dishes [40]. The AR experience at Maxwell Food Centre, for example, gives users an interactive overview of the hawker center's layout and culinary highlights, making it easier for both locals and tourists to navigate and appreciate the diversity of Singapore's food culture [41].

Additionally, Google has mapped the interiors of all 114 hawker centers in Singapore, with each stall assigned a location pin and detailed information accessible through online search. This initiative, conducted in collaboration with the National Heritage Board, National Environment Agency, and the Federation of Merchants' Associations, provides greater online visibility for hawkers and simplifies navigation for visitors, further supporting the digital transformation of Singapore's food markets [42]. The combined effect of digital marketing, AR tourism, and digital mapping has been to modernize Singapore's wet markets and hawker centers, making them more accessible, engaging, and resilient. These innovations have attracted younger, tech-savvy consumers, supported business continuity during disruptions, and positioned Singapore as a global leader in the digital transformation of traditional food markets [27, 41].

3.4 Jeonju Nambu Market, South Korea: Adaptive Reuse and Youth Entrepreneurship

3.4.1 Youth Entrepreneurship Models and Impacts

Jeonju Nambu Market's Youth Mall offers a municipal-backed platform for entrepreneurs under 35, revitalizing previously underused upper floors through creative adaptive reuse [43, 44]. The businesses, ranging from artisanal cafés and cultural merchandise shops to craft breweries and digital content studios, benefit from policies that include subsidized rent and coordinated marketing with the city's tourism bureau. The market's transformation is further supported by a refreshed interior design, curated food experiences, and an active calendar of cultural events such as night markets and live performances, all of which attract younger demographics and foster a vibrant community atmosphere [45].

Strategically located near Jeonju Hanok Village, a major tourist destination, the Youth Mall leverages high visitor flows and integrates seamlessly into the city's broader tourism and urban regeneration strategies [44]. Official tourism materials, local media, and urban policy case studies widely recognize the initiative's integrated spatial design, dynamic programming, and cross-sector partnerships as central to its success [44].

Nonetheless, challenges persist, including seasonal fluctuations in visitor numbers and limited long-term business support for start-ups. Overall, the Jeonju Nambu

Fig. 6 Main gate, Hand-drawn map and the commons at Nambu market Youth mall. Adapted from "전주 남부시장 > 상세 | 관광지·명소 | 열린관광 모두의 여행" [Jeonju Nambu market > details | tourist attractions | accessible tourism for all], Korea tourism organization, in 2024. *Source* Korea tourism organization[2]

Market Youth Mall demonstrates how aligning adaptive physical redevelopment with sustained policy interventions can drive both entrepreneurial vitality and broader market revitalization.

Figure 6 highlights key spatial elements of the Youth Mall, including its redesigned main gate, a hand-drawn wayfinding map, and a communal gathering space that together embody the project's emphasis on accessibility, identity, and collective use.

3.5 London Borough Market, UK: Digital Mapping and Global Engagement

Borough Market in London stands as a leading example of how traditional markets can leverage digital strategies to enhance visitor experience, accessibility, and global engagement. The market has developed a robust digital presence through active engagement on major social media platforms, a dedicated and frequently updated website, and interactive digital mapping tools, all of which have significantly increased its visibility and accessibility for both local and international audiences [31, 46].

A key element of this strategy is Borough Market's strong social media outreach, which enables real-time interaction with followers and has expanded the market's reach far beyond its physical location [31, 47]. As illustrated in Fig. 7, the market's official Instagram page showcases curated visual content that highlights seasonal produce, vendor stories, and behind-the-scenes narratives, reinforcing its cultural brand and digital storytelling approach. The official website acts as a central hub, providing comprehensive information on vendors, events, and visitor guidelines, and its trader-centric design further improves user experience and accessibility [48]. Interactive digital maps, available both online and on-site, help visitors navigate the market's complex layout, locate specific traders, and discover new experiences. This approach has been recognized as a best practice in international benchmarking, credited with improving accessibility and visitor satisfaction [46, 49].

[2] https://access.visitkorea.or.kr/ms/detail.do?cotId=be993a13-58b6-4c51-a98c-29433b2c8a04.

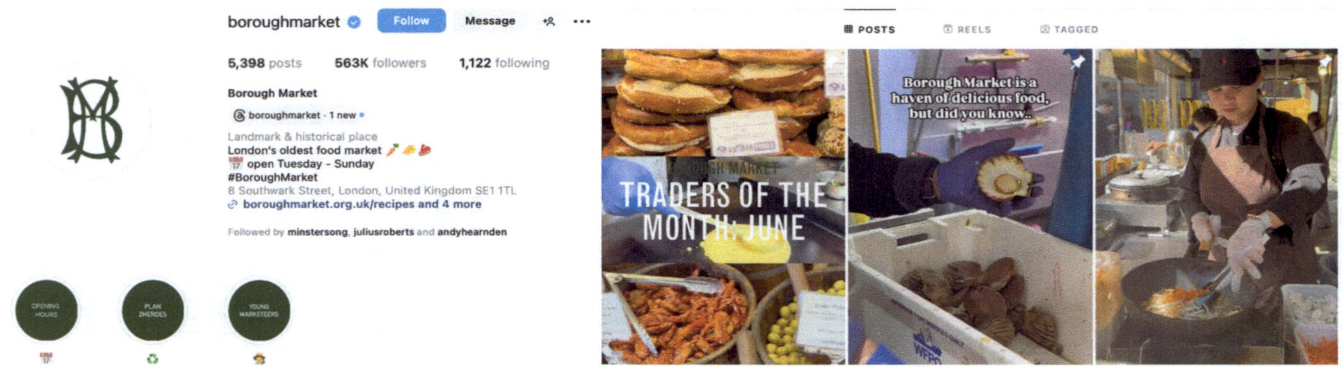

Fig. 7 Borough market official Instagram page, accessed June 20, 2025. *Source* Screenshot from Borough market's official Instagram account

During the COVID-19 pandemic, the market's digital infrastructure proved crucial for maintaining operations and customer engagement. Online platforms were used to communicate safety measures, promote online ordering and delivery, and support vendors, demonstrating the adaptability and resilience of Borough Market's digital strategy [31, 46, 50]. As a result, Borough Market's adoption of digital tools has not only increased global visibility and engagement but also ensured operational continuity during periods of disruption. These achievements have been recognized in tourism and digital innovation reports, positioning Borough Market as a model for other traditional markets seeking to modernize and expand their reach [46].

3.6 *Other International Case Studies*

Across the globe, digital innovation is reshaping traditional food markets through enhanced transparency, traceability, and visitor engagement. Table 1 provides a comparative overview of international case studies, illustrating how digital interventions have shaped various outcomes from enhanced transparency and vendor engagement to expanded market access, operational resilience, and improved visitor experiences.

In the Philippines, platforms like Palengke Boy enable wet market vendors to offer online ordering and contactless delivery, broadening access for urban consumers (Fig. 8a). In China, blockchain-based traceability systems in rural markets allow shoppers to scan QR codes to verify the origin and safety of agricultural products, improving food safety and consumer trust [51].

Globally, technology companies such as EVRYTHNG have partnered with major food brands and retailers, particularly in Europe and North America, to provide Internet of Things (IoT) solutions for supply chain transparency. Their platform creates digital identities for products, enabling real-time tracking and consumer access to sourcing information via QR codes and smart labels [52].

In Italy, the adoption of omnichannel agri-food platforms has enabled small-scale producers to integrate e-commerce with traditional in-person market experiences, thereby expanding their customer reach and strengthening market resilience. These platforms allow producers to leverage digital sales channels alongside physical retail, facilitating broader access to local and regional markets and enhancing the competitiveness of artisanal and smallholder enterprises [53]. This hybrid model has proven particularly valuable in responding to disruptions such as the COVID-19 pandemic, ensuring continuity of supply chains and supporting producer livelihoods.

Similarly, Barcelona has advanced the digital transformation of its food markets by implementing digital mapping tools and a mobile application, Mercats BCN App, that enriches the visitor experience and promotes local food culture. These technologies provide real-time information about vendors, market layouts, and food safety, making market navigation more accessible for both residents and tourists. The integration of digital mapping not only enhances operational efficiency but also

Table 1 Comparative summary of tech-enabled market interventions

Market/Location	Example	Outcomes and impacts of digital integration
Suzhou Shuangta market (China	Digital traceability systems, digital signage, themed retail, social media marketing	Increased visitor numbers, improved transparency, enhanced vendor and community engagement
Hangzhou Wener market (China)	Interactive touchscreens, QR codes for storytelling, mobile app for real-time updates	Enhanced transparency, educational visitor experience, deeper vendor-customer engagement
Singapore wet markets and Hawker centers	Digital marketing (social media, websites), AR tourism, digital mapping, e-payment adoption	Expanded online sales, increased digital engagement, improved navigation, resilience during COVID-19
Jeonju Nambu market (South Korea)	Adaptive reuse (youth mall), modernized interiors, cultural events, integration with tourism	Attracted younger demographic, increased foot traffic, synergy with local tourism
Borough market (UK)	Social media, dedicated website, digital mapping, online ordering, digital communication strategy	Increased global visibility, improved accessibility, operational resilience during disruptions
International (Philippines, Guangzhou China, Barcelona, etc.)	Online ordering platforms (e.g., Palengke Boy), blockchain traceability, IoT supply chain, AR mapping	Broader market access, improved food safety, supply chain transparency, enhanced visitor experience

supports the visibility of local producers and the preservation of culinary heritage [53].

4 Analysis

4.1 Digital Tools as Catalysts for Youth Engagement

Digital technologies such as QR codes, mobile applications, and augmented reality are increasingly reshaping engagement with public heritage and market spaces, especially among younger, tech-savvy consumers who value transparency and convenience [53, 54]. For example, Hangzhou's Wener Market and London's Borough Market have implemented digital displays, QR-linked storytelling, and mobile apps to streamline logistics and enhance consumer trust [31, 36]. In Singapore, Quan Shui Wet Market's adoption of digital marketing and online sales led to a 48% increase in

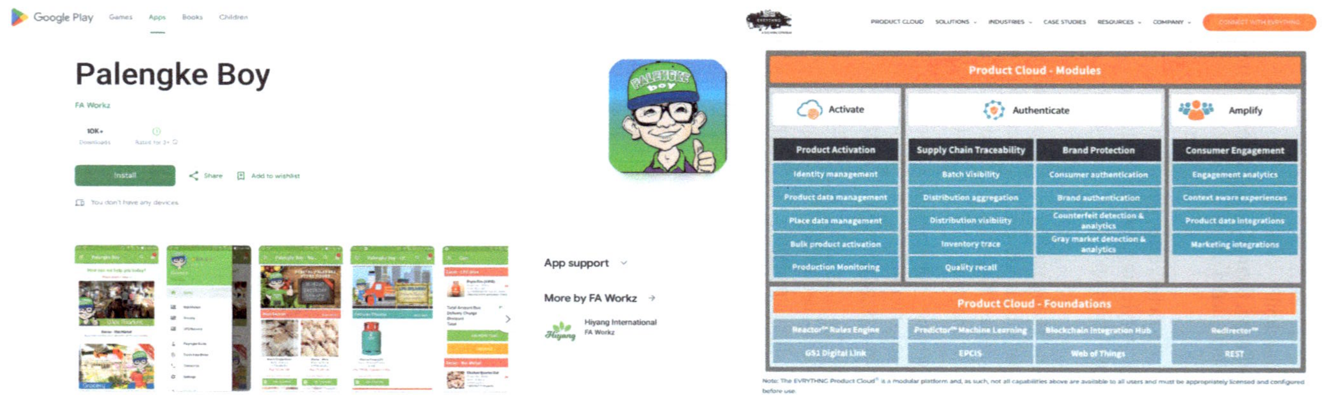

(a) (b)

Fig. 8 **a** Philippines' Palengke Boy app page on google play, accessed June 20, 2025. **b** *EVRYTHNG* product cloud modules and foundations page, accessed June 20, 2025. *Source* Screenshot from EVRYTHNG service description[3]

[3] https://evrythng.com/service-description/.

website visits and now serves 500 households daily, with online sales contributing 40–50% of total revenue [27].

Social media and digital marketing have expanded the reach of traditional markets, as seen in the PRADA x Wuzhong Market collaboration in Shanghai, which attracted large numbers of Gen Z consumers and generated significant public interest through visually driven digital content [55, 56].

4.2 Participatory Design and Adaptive Reuse

Youth participation in co-design and adaptive reuse is recognized as a key factor in revitalizing urban heritage and market spaces [57]. In Jeonju, the creation of the Youth Mall on the second floor of Nambu Market transformed an underutilized space into a vibrant hub for young entrepreneurs, with about 20 stores and regular cultural events, attracting a younger demographic and revitalizing the market atmosphere [45]. This approach has stimulated entrepreneurship and increased community engagement, demonstrating the value of participatory design in sustaining local markets.

4.3 Challenges and Ethical Considerations

While digital campaigns and technology-enabled interventions have generated significant attention and engagement for wet markets, they also introduce a range of challenges and ethical concerns. For instance, the PRADA x Wuzhong Market collaboration in Shanghai successfully attracted Gen Z consumers and generated global media coverage, but it also raised concerns about the reduction of local culture to spectacle and the amplification of commercial gentrification [55, 56, 58]. Research on geomedia and digital placemaking highlights how such interventions can magnify the influence of dominant social groups, potentially marginalizing traditional vendors and long-standing community members in favor of affluent or tourist-oriented users [8, 58]. This process, often described as digital or commercial gentrification, risks transforming markets into curated destinations that prioritize visual appeal and consumer experience over local authenticity and social equity.

Another major challenge is ensuring digital accessibility and inclusion. While digital tools can enhance market transparency and engagement, they may inadvertently exclude elderly populations, low-income groups, or those with limited digital literacy [59, 60]. Reports highlight that digital divides persist, particularly affecting youth, women, and marginalized groups, and emphasize the need for accessible technologies and digital skills training to ensure no one is left behind [59, 60]. Data privacy and security are also critical concerns, as the increasing use of digital payment systems, traceability apps, and online engagement platforms requires robust protections for personal and transactional data [61]. The European Commission underscores the importance of strong data protection frameworks, such as the GDPR, to safeguard users as digitalization expands across sectors [61]. Without careful attention to these issues, digital transformation may reinforce existing inequalities and undermine the social fabric that has historically defined wet markets. Addressing these challenges requires a commitment to equitable design, ongoing community consultation, and the development of inclusive digital infrastructures that serve all market stakeholders.

4.4 Governance and Participation

The vitality and resilience of wet markets as urban commons depend on inclusive governance structures and meaningful participatory platforms. As demonstrated in the case studies, markets thrive when local stakeholders, including youth, vendors, residents, and civil society organizations, are actively involved in both digital and physical aspects of market life [28, 43]. Best practices in participatory governance emphasize the importance of shared decision-making, transparent management, and collaborative storytelling, which together foster a sense of ownership and long-term sustainability [8, 62, 63].

Recent research on placemaking governance suggests that municipal authorities and market managers should adopt adaptive, facilitative roles, acting as "intelligent hosts" who support bottom-up initiatives and mediate between competing place narratives [62, 63]. This approach encourages the co-production of market space, where diverse voices, including those of youth and marginalized groups, are integrated into planning, programming, and digital innovation [23, 28]. Digital participatory tools, such as online feedback platforms and co-design workshops, can further democratize governance processes and ensure that market revitalization reflects the needs and aspirations of the whole community [9, 28, 64]. Ultimately, effective governance balances innovation with tradition, enabling wet markets to adapt to contemporary challenges while sustaining their cultural and social roles.

5 Discussion

The comparative case studies presented in this chapter reveal that the digital transformation of wet markets is most successful when it is grounded in inclusive, participatory, and context-sensitive strategies. As seen in Suzhou's Shuangta Market and Hangzhou's Wener Market, the thoughtful integration of digital infrastructure, such as QR code-based traceability and interactive digital screen and kiosk, has enhanced transparency, operational efficiency, and visitor engagement, while also supporting the retention of original vendors and the preservation of local heritage [32, 35–37]. In Singapore, the combination of digital marketing, augmented reality tourism, and comprehensive digital mapping has modernized wet markets and hawker centers, attracting younger, tech-savvy consumers and supporting business continuity during disruptions such as the COVID-19 pandemic [27, 41].

The Jeonju Nambu Market case highlights the value of adaptive reuse and youth entrepreneurship, demonstrating how underutilized spaces can be transformed into vibrant hubs for creative business and cultural activity [43, 44]. Borough Market in London illustrates how robust digital communication and mapping strategies can expand global reach, improve accessibility, and ensure operational resilience during crises [31, 46]. However, the analysis also underscores the risks of digital gentrification and exclusion, particularly when interventions prioritize spectacle or commercial interests over local participation and equity [56, 58].

The findings suggest that the most resilient and vibrant markets are those that combine digital innovation with participatory governance, intergenerational collaboration, and a strong commitment to cultural continuity. Youth involvement in co-design and storytelling, as seen in Jeonju and Seoul, is especially important for bridging generational divides and sustaining market relevance [28, 57]. Institutional frameworks that support shared decision-making, digital literacy, and ongoing community engagement are critical for ensuring that digital transformation enhances, rather than erodes, the social and cultural value of wet markets. By learning from these diverse cases, policymakers and practitioners can develop scalable models for market revitalization that balance innovation with heritage and prioritize the needs of all stakeholders.

6 Policy and Future Research Implications

We argue that revitalizing traditional wet markets in the digital age requires a comprehensive policy approach that balances technological innovation with the preservation of cultural authenticity. To promote digital equity, we recommend making digital tools accessible to all stakeholders, including elderly vendors, low-income patrons, and marginalized groups, through subsidized infrastructure, targeted outreach, and digital literacy programs. To prevent digital gentrification, we emphasize the importance of retaining traditional vendors, supporting affordable stall rents, and fostering community-led digital initiatives that maintain the social fabric of markets.

We contend that sustainable transformation depends on inclusive governance. By institutionalizing participatory models, such as co-creation workshops, digital feedback platforms, and regular stakeholder consultations, we ensure broad engagement in decision-making. We also advocate for updated regulatory frameworks that address challenges like platform dominance, data privacy, and fair competition. Ongoing investment in digital literacy and user-friendly platforms remains essential to bridging digital divides, while modernization efforts must safeguard intangible cultural heritage.

For future research, we propose conducting longitudinal studies to assess the social impact of digital interventions, developing robust metrics for digital equity, and undertaking comparative analysis of digital gentrification. Experimenting with participatory design and governance, and piloting cross-city collaborations, will help us identify scalable, inclusive strategies for revitalizing wet markets.

Policy and Research Proposals

- Promote digital equity with subsidized infrastructure and digital literacy training.
- Prevent digital gentrification by supporting traditional vendors and affordable rents.
- Institutionalize participatory governance through co-creation workshops and digital feedback.
- Update regulations for data privacy, platform power, and fair competition.
- Invest in outreach and user-friendly digital platforms to bridge divides.
- Safeguard intangible cultural heritage alongside modernization.
- Conduct longitudinal and comparative studies on digital interventions and equity.
- Pilot cross-city collaborations for scalable, innovative revitalization.

7 A Scalable Framework Proposal

The proposed Framework (Fig. 9) for Revitalizing Wet Markets, illustrated in the attached diagram, synthesizes key insights from diverse case studies into a holistic, scalable strategy for transforming wet markets into vibrant urban commons. Drawing on the commons-based governance model, the framework positions wet markets as collectively managed resources, emphasizing inclusivity, shared responsibility, and co-creation [8].

At the heart of the framework lies **Cultural Continuity**, which protects and transmits local foodways, social rituals, and place-based knowledge. This is essential for maintaining the identity of markets while enabling modernization. Studies by Li et al. and Zhong et al. confirm that grounding revitalization efforts in cultural practices ensures community relevance and long-term engagement [6, 65].

Civic Participation is a central pillar, operationalized through youth engagement and vendor involvement. Initiatives such as Jeonju's Youth Mall and Singapore's vendor-led digital strategies highlight how inclusive engagement fosters market vibrancy [27, 57].

Digital Placemaking tools, including QR codes, mobile apps, and digital storytelling, enhance market accessibility and transparency while facilitating intergenerational connections [9, 53]. These tools also support market traceability and visitor experience enhancement.

Participatory Design reinforces these efforts by transforming market spaces into youth-led hubs and cultural venues. Adaptive reuse strategies, exemplified by initiatives in East Asia, ensure spatial resilience and relevance in changing urban contexts [43, 57].

Effective Governance involves establishing Market Councils with diverse representation and providing Digital Literacy Training. These steps build the institutional capacity needed for sustained community-driven market management [26, 27].

To ensure **Equity and Sustainability**, the framework includes subsidized digital infrastructure and data privacy protections to prevent exclusion and mitigate digital

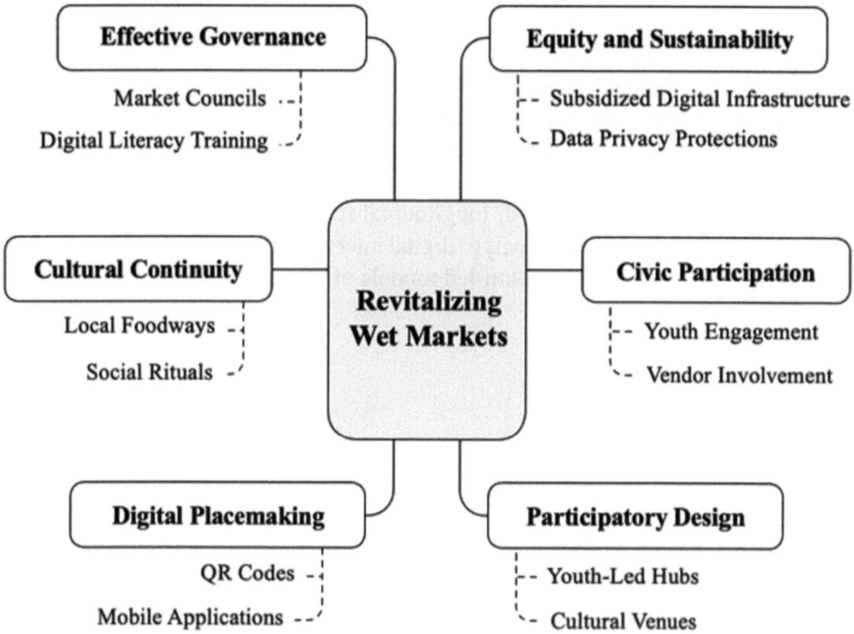

Fig. 9 Framework for revitalizing wet markets

gentrification [53, 61]. These measures safeguard vulnerable populations while fostering innovation.

Crucially, the framework is adaptive and iterative, encouraging longitudinal impact assessments, stakeholder feedback loops, and best practice documentation. This ensures that the revitalization process remains responsive to local needs and replicable in diverse contexts. By integrating cultural, civic, digital, and governance strategies, this framework positions wet markets as resilient, inclusive, and dynamic urban nodes, prepared to thrive in the digital era while rooted in tradition and community life.

8 Conclusion

This chapter has argued that revitalizing traditional wet markets in the digital age requires a holistic approach rooted in cultural continuity, intergenerational engagement, and participatory governance. Through case studies across East Asia and beyond, it has demonstrated how digital placemaking tools, QR codes, mobile apps, AR, and digital storytelling can foster transparency, traceability, and youth engagement while respecting the social and spatial fabric of markets. The integration of digital strategies with co-design, adaptive reuse, and inclusive governance has been shown to increase footfall, strengthen community identity, and preserve intangible heritage. However, these innovations are not without challenges: digital gentrification, uneven accessibility, and the risk of eroding authenticity highlight the need for ethical design and inclusive policy frameworks. Markets that succeed in this transformation are those that empower both younger and older generations, prioritize equity, and blend tradition with innovation in context-sensitive ways. To guide future urban interventions, this study offers a scalable framework that centers markets as vibrant urban commons. Moving forward, longitudinal research is needed to measure the long-term social and cultural impacts of digital interventions, with particular attention to cross-city comparisons and youth-led models of governance that sustain markets as inclusive civic spaces.

References

1. Krausz M, Strehlau V, Choi F, Jang KL (2019) Chapter 8 urbanization and marginalization. In: Urban mental health. OUP Oxford, United Kingdom
2. Wang Y, Coe NM (2021) Platform ecosystems and digital innovation in food retailing: exploring the rise of Hema in China. Geoforum 126:310–321. https://doi.org/10.1016/j.geoforum.2021.08.007
3. Deloitte (2023) Gen Z and millennial survey
4. Zhong S (2022) Saving China's wet markets. Sixth Tone
5. El Shandidy MZ (2023) The power of intangible heritage in sustainable development. Power 6:92–97
6. Zhong S, Crang M, Zeng G (2020) Constructing freshness: the vitality of wet markets in urban China. Agric Hum Values 37:175–185. https://doi.org/10.1007/s10460-019-09987-2
7. Zhong T, Si Z, Crush J, Xu Z, Huang X, Scott S, Tang S, Zhang X (2018) The impact of proximity to wet markets and supermarkets on household dietary diversity in Nanjing City. China. Sustainability 10(5):1465. https://doi.org/10.3390/su10051465
8. Foster S, Iaione C (2016) The city as a commons. Yale Law Policy Rev 34:281. https://doi.org/10.2139/ssrn.2653084
9. Foth M, Hudson-Smith A, Gifford D (2016) Smart cities, social capital and citizens at play: A critique and a way forward. In: Foth M, Brynskov M, Ojala T (eds) Citizen's right to the digital city: urban interfaces, activism, and placemaking. Edward Elgar Publishing, pp 205–221. https://doi.org/10.4337/9781784717766.00017
10. Lovendianto J, Samodra FT, Ekasiwi SNN, Kim DY (2025) Digitally enhanced architecture for sustainable placemaking: a systematic review. IOP Conf Ser: Earth Environ Sci IOP Publishing 1439(1):012025). https://doi.org/10.1088/1755-1315/1439/1/012025
11. Downs SM, Ahmed S, Warne T, Fanzo J, Loucks K (2022) The global food environment transition based on the socio-demographic index. Glob Food Sec 33:100632. https://doi.org/10.1016/j.gfs.2022.100632
12. Wertheim-Heck SCO, Raneri JE (2020) Food policy and the unruliness of consumption: an intergenerational social practice approach to uncover transforming food consumption in modernizing Hanoi, Vietnam. Glob Food Secur 26:100418. https://doi.org/10.1016/j.gfs.2020.100418
13. Bharucha ZP, Weinstein N, Watson D, Boehm S (2020) Participation in local food projects is associated with better psychological well-being: evidence from the east of England. J Public Health 42(2):e187e197. https://doi.org/10.1093/pubmed/fdz057
14. South China Morning Post (2018) Wet markets face survival battle

15. Geng X-H, Chen K, Wei A, Zhou Y (2017) Market competition of fresh vegetables between supermarket and wet market in China based on the difference of quality and price. In: Proceedings of the 9th Asian society of agricultural economists (ASAE) international conference, pp 1–23. https://doi.org/10.22004/ag.econ.284876
16. Wertheim-Heck SCO, Vellema S, Spaargaren G (2015) Food safety and urban food markets in Vietnam: the need for flexible and customized retail modernization policies. Food Policy 54:95–106. https://doi.org/10.1016/j.foodpol.2015.05.002
17. Chowdhury S, Das TK, Chisty NN, Biswas SK, Rahman MZ, Ling J, Naguib MM, Lindahl JF (2025) Regulations, biosecurity measures, and impact of COVID-19: a comprehensive mixed method study in traditional wet and live animal markets in Bangladesh. One Health 20:101014. https://doi.org/10.1016/j.onehlt.2025.101014
18. Koberinski J, Si Z, Scott S (2019) The food safety and food security nexus in the urbanizing Global South (Hungry Cities Partnership Discussion Paper No. 35). Hungry Cities Partnership
19. Hofman M, Trevenen-Jones A (2024) Wet markets in Southeast Asia and access to healthy diets. Front Sustain 5, Article 1320959. https://doi.org/10.3389/frsus.2024.1320959
20. Daya T, Taylor M (2021) Improving wet market regulation to control the spread of disease. Asian-Pac Law Policy J 23(1):97–125
21. Heron G, Steckley L (2018) Digital storytelling using co-production with vulnerable young people. J Soc Work 20(4):411–430. https://doi.org/10.1177/1468017318814993. Original work published 2020
22. The British Academy (2021) Digital placemaking: Creating synergies between digital and economic recovery and health, care and well-being strategies
23. Tsang YF, Chan SMH, van Ameijde J (2024) Digitally gamified co-creation: enhancing community engagement in urban design through a participant-centric framework. https://doi.org/10.1017/dsj.2024.17
24. Bridging generational gaps in cultural preservation: Engaging youth with heritage through digital innovation (2025) LinkedIn Articles
25. Heritage Trust Network Youth Forum (2024) Placemaking manifesto digital version
26. Wai-Kit Ng C-L, Chen C-L, Huang Y-H (2024) Revitalization of cultural heritage in the digital era: a case study in Taiwan. Urban Resil Sustain 2(3):215–235. https://doi.org/10.3934/urs.2024011
27. Enterprise Singapore (2025) Quan Shui Wet Market
28. UN-Habitat (2024) Digital tools for participatory urban planning led by youth
29. Shih C-M, Treija S, Zaleckis K, Bratuškins U, Chen C-H, Chen Y-H, Chiang CTW, Jankauskaitė-Jurevičienė L, Kamičaitytė J, Koroļova A, Lee H-C, Lektauers A, Mlinkauskienė A (2022) Digital placemaking for urban regeneration: Identification of historic heritage values in Taiwan and the Baltic States. Urban Plan 7(3):302–315. https://doi.org/10.17645/up.v7i3.4406
30. Djukic A, Stober D, Tiano P, Negru M, Maric J, Sepe M, Economou A (2024) Placemaking and networking of heritage for sustainable tourism. In: Placemaking in practice 1:11–36. Brill. https://doi.org/10.1163/9789004542389_004
31. Borough Market (2022) Report and financial statements for the year ended 31 March 2022. Borough Market Trust
32. China Daily (2021) Suzhou makes best use of ancient and modern
33. China Daily (2023) Farmers' markets sow seeds for tourism boom
34. GS1 (2023) Better food safety and supply chain efficiency in Zhejiang Province, China: 2D barcodes case study
35. 央视网(CCTV.com), "老菜场焕新逆袭链起城市记忆与市井'烟火气' 消费新场景释放经济新活力(The old market has been rejuvenated and made a comeback, connecting the city's memories with the "vitality of daily life" of the market. New consumption scenarios have released new economic vitality)," 央视网 (2025). https://news.cctv.com/2025/05/23/ARTI91jDH8b5GgJC2yM0dltL250523.shtml. Last accessed 1 Jun 2025
36. Hangzhou (2024) Wen'er market: Hangzhou's trendy fresh market blends tradition, modernity

37. 央视网(CCTV.com), "[消费主张]杭州文二菜市 不仅品质好服务更是佳([Consumption Claim] Hangzhou Wen Er Market not only offers excellent quality but also provides superior service)," 央视网 (2025). https://tv.cctv.com/2025/01/08/VIDEB9wPb4rEKwbg2rcjgt6m250108.shtml.
38. Ektha (2025) QR codes in 2025: How the latest trends are reshaping customer engagement. In: State of QR Codes 2025 Report. Uniqode
39. The Straits Times (2024) How hawker brothers are shaking up family's wet market business. https://www.straitstimes.com/business/companies-markets/hawker-brothers-digitalising-modernising-family-wet-market-business-quan-shui-wet-market-enterprise. Last accessed 1 June 2025
40. FoodSided (2024) Google maps AR offers the ultimate walk-through Singapore's Maxwell food centre
41. Google (2024) Discover Singapore through an immersive augmented reality tour
42. NEA (National Environment Agency) (2019) Google puts Singapore hawker centres on the map
43. Jeonju City Government and UCLG ASPAC (2023) 2023 Jeonju Future Cities Forum: Building the future and promotion of the youth
44. Korea Tourism Organization (2024) Jeonju Nambu Traditional Market (전주 남부시장). Visit Korea. Last accessed 30 June 2025
45. Jeonju City (2024) Youth mall at Nambu market. Visit Jeonju. Last accessed 30 June 2025
46. Fáilte Ireland (2021) Benchmarking international best practice in markets. Fáilte Ireland
47. Hampson L (2020) Borough Market is posting live cookery lessons on Instagram. Time Out London
48. Borough Market (2024) Annual report and financial statements for the year ended 31 March 2024. Borough Market Trust
49. O'Neill, J. (2013) Borough Market launches new interactive map. Londonist
50. Barley Communications (2024). Communicating in the times of coronavirus with Borough Market
51. Ledger Insights. (2020). China's food markets use blockchain for traceability, consumer safety.
52. Cointelegraph (2019) Iota, IoT firm Evrythng partner to increase transparency for consumer goods supply chain
53. Food and Agriculture Organization (2022b) Socially inclusive digital tools for agriculture
54. Stripe (2023) QR codes explained: How they're used for payments
55. European Guanxi (2024) How Prada engages with Chinese Millennials through digital communication
56. South China Morning Post (2021) Prada takes over Shanghai wet market to tap Gen Z consumers
57. Zhang Y, Ikiz Kaya D, van Wesemael PJV, Colenbrander BJF (2024) Youth participation in cultural heritage management: a conceptual framework. Int J Herit Stud 30(1):56–80. https://doi.org/10.1080/13527258.2023.2275261
58. Xie Y, Chen C, Zhao Y, Deng J (2024) The amplification of commercial gentrification in internet-famous places by geo-media: the case of Wuzhong Market in Shanghai. https://doi.org/10.2139/ssrn.4973838
59. Dittmer KM, Wollenberg E, Burns SA (2022) Socially inclusive digital tools for agriculture: a way forward. CGIAR
60. Food and Agriculture Organization (2022a) ICTs for well-being, inclusion and resilience: through digital skills for youth and women in agriculture
61. European Commission (2025) Data protection in the EU
62. Meerkerk J (2024) Understanding the governing of urban commons: reflecting on five key features of collaborative governance in Zero Waste Lab, Amsterdam. Int J Commons 18(1):397–410. https://doi.org/10.5334/ijc.1282
63. Wolfram F (2021) The ABC's of placemaking governance: learning from Amsterdam, Berlin and Copenhagen (Master's thesis, KTH Royal Institute of Technology). DiVA portal
64. Dittrich Y, Ddamba J (2015) Data for participation and participation as data: supporting incremental participatory decision-making in urban planning. IRIS: Selected Papers of the Information Systems Research Seminar in Scandinavia 6:16

65. Li C, Cui Z, Chen J, Zhou N (2019) Brand revitalization of heritage enterprises for cultural sustainability in the digital era: a case study in China. Sustainability 11(6):1769. https://doi.org/10.3390/su11061769

Contextual Applications

Digital Democratization of Heritage: Participatory Platforms for Revitalization of Historic Sites

Ahmet Kurnaz

Abstract This study examines how digital participatory platforms influence urban heritage revitalization in historic cities, emphasizing cases from Madrid, Helsinki, Reykjavik, Amsterdam, and Barcelona. Traditional top-down governance often limited community input, hindering project sustainability and acceptance. To address these gaps, contemporary approaches use digital tools (e.g., mobile apps, GIS, online forums) to foster direct citizen engagement and transparent decision-making. Analysis reveals platforms such as "Decide Madrid," "FixMyStreet," and "Maptionnaire" improve citizen participation and transparency. Moreover, "Amsterdam Smart City" and "Decidim" demonstrate substantial entrepreneurial impacts through CivicTech startups, enhancing economic and cultural outcomes. However, significant challenges persist, including digital divides, platform usability issues, and data privacy concerns. The study recommends integrated digital-traditional participation models, improved platform design, and targeted digital literacy initiatives to maximize effectiveness and equity.

Keywords Digital democratization · Urban heritage revitalization · Participatory platforms · Civictech entrepreneurship · Inclusive urban governance · Digital divide

1 Introduction

The integration of digital technologies in urban planning and heritage conservation has emerged as a transformative phenomenon in recent decades, significantly reshaping conventional urban management paradigms [1]. Particularly in historic neighborhoods, where urban heritage intersects with contemporary socio-economic

A. Kurnaz (✉)
Istanbul Sabahattin Zaim University, Istanbul, Turkey
e-mail: ahmet.kurnaz@izu.edu.tr

V. Chand Sharma et al. (eds.), *Tech-Enabled Urbanism and Entrepreneurship for Inclusive Cities*, Lecture Notes in Civil Engineering 779,
https://doi.org/10.1007/978-981-95-3873-7_10

dynamics, the potential for digital tools to foster inclusive and sustainable urban revitalization is increasingly evident [2]. Traditional approaches to heritage conservation have followed top-down methodologies, characterized by centralized decision-making processes and limited community engagement [3]. This hierarchical model has often resulted in a disconnection between local communities and urban revitalization initiatives, weakening the long-term sustainability and social acceptance of heritage projects [4].

Such digital platforms empower communities by providing concrete mechanisms for direct citizen input into heritage management processes, such as GIS-based reporting tools and interactive decision-making forums. This study uniquely investigates how platforms like "Decide Madrid" and "Maptionnaire" actively enable diverse community groups to directly influence specific heritage revitalization projects rather than merely consulting them. Additionally, the research explores how these platforms support entrepreneurial activities linked explicitly to heritage assets, thus contributing to both democratic and economic revitalization goals simultaneously [5–7].

Simultaneously, the use of digital participatory tools extends beyond mere community engagement, serving as catalysts for entrepreneurial innovation within urban contexts. CivicTech startups, defined as entrepreneurial ventures that address urban issues through technology, have become pivotal actors in leveraging digital democracy to stimulate local economic activities and entrepreneurial solutions [8, 9]. Through accessible digital platforms, these startups can effectively identify community needs, develop responsive services, and foster local entrepreneurship aligned with heritage conservation and urban revitalization goals [10].

However, despite their evident benefits, digital technologies in heritage neighborhoods also present notable challenges. The digital divide issue—the unequal access to technology among community members—remains a significant barrier, potentially reinforcing existing social inequalities rather than alleviating them [11]. Additionally, concerns regarding the superficial nature of participation, data privacy, platform usability, and the authenticity of digitally mediated interactions highlight critical limitations that need strategic consideration [12].

Given this complex context, this research explores the transformative potential of digital participatory platforms in enhancing community engagement, fostering entrepreneurial activities, and overcoming existing socio-technical limitations within historic neighborhoods. The research aims to provide theoretical insights and practical guidance for urban planners, policymakers, heritage conservationists, and social entrepreneurs through comprehensive case analyses and stakeholder perspectives. By investigating these dimensions, this study contributes to the burgeoning literature on digital urbanism and social entrepreneurship and offers actionable strategies for creating more inclusive, sustainable, and vibrant urban ecosystems.

However, despite a growing body of research on digital urbanism and entrepreneurial urban solutions, there remains a distinct research gap concerning the intersection of digital democracy, entrepreneurial urbanism, and heritage revitalization. Specifically, previous studies have often treated these areas independently or only partially integrated them. The novelty of this study lies in explicitly bridging this gap by examining how digital participatory platforms simultaneously enhance democratic engagement and entrepreneurial activities, specifically within heritage contexts, thereby addressing a critical void in understanding how these platforms can be operationalized to achieve holistic urban sustainability and economic revitalization goals.

2 Context and Literature Review

2.1 Urban Heritage Revitalization and Social Inclusivity

Urban heritage revitalization refers to the complex process of reintegrating historic built environments into contemporary urban life while preserving their cultural and historical significance [4]. Historically, such efforts have emphasized aesthetic preservation and architectural integrity, often at the expense of social and economic inclusion [13]. However, recent discourse has shifted toward recognizing heritage sites as dynamic socio-cultural assets that can contribute to inclusive and sustainable urban development [14].

In this context, social inclusivity ensures that diverse community groups, regardless of socioeconomic status, ethnicity, or age, have equitable access to the benefits and opportunities generated through heritage-led revitalization. Studies indicate that inclusive planning processes enhance local support, cultural continuity, and collective memory, all of which are critical to the success of revitalization initiatives [15]. Moreover, inclusive urban heritage practices align with the principles of the New Urban Agenda, which emphasize participatory governance, cultural sustainability, and community-led development [16].

However, despite the inclusive rhetoric surrounding urban heritage revitalization, critical debates persist concerning whether these approaches benefit all community groups equally. Critics argue that heritage-led regeneration projects often lead to gentrification, inadvertently displacing economically vulnerable populations despite intentions of inclusivity [4, 13]. Additionally, scholars highlight tensions between preserving architectural integrity and addressing contemporary socio-economic needs, emphasizing that heritage conservation sometimes prioritizes aesthetic considerations over community functionality and sustainability [15, 17].

2.2 Digital Democracy and Urban Planning

Digital democracy, also referred to as e-democracy or electronic participation (e-participation), encompasses the use of digital tools to enhance citizen involvement in public decision-making processes [18]. Within urban planning, digital democracy seeks to create transparent, responsive, and participatory planning frameworks that bridge the gap between authorities and communities [19].

A growing body of research highlights the transformative role of platforms such as online forums, crowdsourcing portals, and participatory mapping systems in reshaping urban governance. For example, GIS-based platforms like "FixMyStreet" and "Maptionnaire" have allowed residents to report issues and co-produce urban knowledge, thus fostering a sense of agency and ownership [20]. Mobile applications further democratize access by enabling real-time feedback and localized participation, particularly in underserved communities where traditional public engagement mechanisms are less effective [21].

Despite these advantages, scholars have also noted that digital participatory platforms can inadvertently reproduce patterns of exclusion, particularly if access to digital infrastructure and digital literacy is not equitably distributed [22]. Thus, effective implementation requires hybrid approaches that combine digital tools with offline engagement strategies to ensure broad accessibility and inclusivity.

Nevertheless, despite their transformative potential, digital participatory platforms have faced considerable criticism. For instance, platforms like "FixMyStreet" have sometimes been accused of promoting a tokenistic form of participation, where citizens' reports are acknowledged but rarely lead to substantial policy changes or improvements in local governance [22, 23]. Furthermore, research indicates that unequal digital access and varying digital literacy levels significantly restrict the representativeness of participation, raising concerns that digital tools might reinforce existing inequalities rather than mitigate them [11, 21].

2.3 Entrepreneurial Urban Solutions and CivicTech Startups

Entrepreneurial urbanism refers to integrating entrepreneurial strategies within urban governance to foster innovation, adaptability, and economic development [24, 25]. In recent years, CivicTech startups—entrepreneurial ventures leveraging digital technologies for civic improvement—have emerged as key actors in this space, particularly within urban heritage contexts.

These startups utilize open data, digital platforms, and participatory technologies to co-create urban services and solve community-identified problems [26]. Examples include apps for community-led urban design, heritage mapping, and micro-investment platforms for restoring cultural assets [27]. Such initiatives enhance transparency and engagement and create new forms of socioeconomic value, contributing to local entrepreneurship and job creation [28].

However, the sustainability and scalability of CivicTech solutions depend on policy support, cross-sector collaboration, and long-term digital infrastructure investments [29]. Moreover, as entrepreneurial models become embedded in urban planning, concerns arise about commercializing civic spaces and the risk of privileging tech-savvy elites [30]. Addressing these tensions requires a critical understanding of the socio-political implications of digital entrepreneurship in heritage contexts.

Despite the positive impacts of CivicTech startups, significant critical debates exist regarding their implications for urban governance and social equity. Critics argue that entrepreneurial urbanism, while innovative, risks commercializing civic functions and public spaces, potentially privileging wealthier and tech-savvy populations at the expense of broader community needs [28, 30]. Additionally, scholars caution that relying on entrepreneurial models for urban governance might shift responsibility away from the public sector, leading to inconsistencies in service delivery and amplifying socio-economic disparities [25, 29].

3 Methodology

3.1 Research Design and Case Study Selection

This research employs a comparative case study methodology, a practical approach in exploring complex phenomena within real-life contexts [31]. Comparative case analysis allows for an in-depth examination of how digital participatory platforms function across different socio-cultural and economic contexts. It offers insights into commonalities and distinctive features influencing their success or failure. Primary contexts were selected: historic neighborhoods European cities, chosen due to their rich heritage, varying socio-economic structures, and ongoing digital participation initiatives.

The case selection criteria included the explicit use of active digital participatory platforms in heritage conservation and urban revitalization, alongside diversity in scale, stakeholder involvement, and entrepreneurial outcomes. Each city specifically aligns with these criteria as follows:

- Madrid was selected due to its innovative and extensively studied platform "Decide Madrid," known for substantial community involvement and direct democratic participation in heritage-related urban planning [7].
- Helsinki was chosen for the advanced use of GIS-based participatory planning through "Maptionnaire," extensively recognized for inclusive and effective urban heritage management [6].
- Reykjavik's inclusion is based on the successful civic engagement and urban policymaking results documented with the platform "Better Reykjavik," which explicitly addresses community-identified heritage revitalization projects [32].
- Amsterdam provides a critical case through the "Amsterdam Smart City" initiative, emphasizing collaboration between municipalities, startups, and citizens, effectively fostering entrepreneurial activities directly tied to heritage conservation [33].
- Barcelona was included due to its pioneering open-source platform "Decidim," known for explicitly stimulating significant entrepreneurial impacts and socio-economic regeneration within heritage districts [34].

- Lisbon, Milan, and London are grouped under the "Sharing Cities" initiative, selected to illustrate how coordinated multi-city digital strategies can amplify entrepreneurial and civic outcomes in heritage contexts, providing a valuable comparative dimension [35].

3.2 Data Collection Methods

Data collection involved a mixed-method approach, combining qualitative and quantitative data sources to ensure methodological triangulation and enhance research validity [36].

Stakeholder Interviews: Semi-structured interviews were conducted with a diverse range of stakeholders. Participants were selected to ensure varied perspectives and expertise [37]. Interviews focused on stakeholders' experiences, perceived effectiveness of digital platforms, entrepreneurial activities stimulated by digital participation, and barriers encountered.

Platform Data Analysis: Quantitative data analyzed from the digital participatory platforms included detailed user analytics collected directly from platform operators and respective municipalities. The collected data encompassed user registration counts, frequency of platform usage, types of user interactions (e.g., votes, comments, issue reporting), geographic distribution of participants, and engagement duration. Specific attention was given to evaluating the platforms' efficacy in addressing community-generated heritage concerns and their direct influence on urban planning decisions, thus providing comprehensive insights beyond general participation rates.

3.3 Analytical Framework and Evaluation Metrics

The analytical framework integrates qualitative content analysis with descriptive statistical methods. Thematic content analysis was conducted on interview transcripts, following Braun and Clarke's guidelines to identify recurring themes related to community participation, entrepreneurial outcomes, and platform usability [38].

Quantitative data were analyzed using descriptive statistics, focusing on user participation rates, and types of entrepreneurial initiatives launched. The effectiveness of digital platforms was evaluated using an adaptation of Arnstein's Ladder of Participation, assessing levels of citizen empowerment and decision-making influence achieved through digital interactions [39].

Metrics explicitly developed for this research included:

- User participation frequency and diversity: Frequency was measured by calculating monthly active users and average session duration from platform analytics.

Diversity was quantitatively assessed through demographic profiling data and geographic distribution data provided by municipalities and platform operators.
- Entrepreneurial activities initiated through platform engagement: This metric was operationalized by counting and categorizing new CivicTech startups launched, the number of jobs created, and total investments secured, using data from municipal entrepreneurship offices and startup accelerators.
- Comparative analysis of platform impact on community engagement and heritage project legitimacy across cases: Impact assessment involved pre- and post-intervention metrics, including changes in citizen-reported satisfaction levels, frequency of heritage-related project initiations, and perceived legitimacy scores gathered through stakeholder surveys and municipal reports.

This rigorous methodological framework ensures comprehensive insights into the role and effectiveness of digital participatory platforms, guiding future practices and policies for sustainable and inclusive heritage-led urban revitalization.

4 Results and Case Studies

4.1 Case Study Analysis: Digital Platforms Enhancing Community Participation

A comprehensive examination of several prominent digital platforms highlights their substantial role in enhancing community participation within heritage-focused urban revitalization initiatives. One notable case is Madrid's participatory platform, "Decide Madrid." Initiated by Madrid's city council, this platform facilitates public engagement by allowing citizens to propose, debate, and vote on urban projects, including heritage conservation and urban development plans [7]. Detailed analyses of platform data illustrate that Decide Madrid has engaged over 400,000 registered users actively involved in decision-making, resulting in tangible contributions to heritage regeneration projects such as the revitalization of Plaza de España. Stakeholders indicated improved transparency, legitimacy, and increased public trust due to open-access decision-making processes and direct citizen influence on urban policy outcomes (Table 1). This observation explicitly emerged from a thematic analysis of stakeholder interviews conducted for this research.

Similarly, "FixMyStreet," a UK-based digital participatory tool, significantly enhances urban governance by enabling direct citizen-to-government communication. This platform has recorded over one million citizen-generated reports annually, empowering residents to promptly report infrastructure issues, including concerns related to heritage preservation [23]. According to municipal records, the timely response facilitated by this digital intervention has notably improved citizen satisfaction, urban service efficiency, and engagement in civic processes.

Table 1 Comparative overview of selected digital participatory platforms

Platform name	Location	Main purpose	Technology used	User engagement metrics	Impact on heritage conservation
Decide Madrid	Madrid, Spain	Participatory decision-making	Online Forums, Voting Systems	400,000 + active users	Enhanced decision-making legitimacy
FixMyStreet	UK (multiple cities)	Citizen-to-government communication	Mobile App, GIS	1 million + annual reports	Improved heritage infrastructure reporting
Maptionnaire	Helsinki, Finland	Participatory urban planning and GIS	GIS, Interactive surveys	250,000 + active respondents across planning projects	Improved urban planning and heritage management
Better Reykjavik	Reykjavik, Iceland	Collaborative urban policymaking	Web Platform, Voting System	Over 30,000 registered users contributing annually	Prioritized heritage conservation projects

Source By Authors

Helsinki's "Maptionnaire," another exemplary initiative, provides insights into the participatory use of GIS (Geographic Information Systems) for community-driven urban planning and heritage preservation. Through this interactive platform, community members actively contribute spatially explicit insights, significantly informing urban planners and enhancing the democratic legitimacy of urban master plans [6]. Detailed metrics and stakeholder evaluations consistently reported high levels of user participation, highlighting the platform's success in integrating diverse community voices into heritage management and urban policy formulation.

Furthermore, the "Better Reykjavik" platform exemplifies effective citizen engagement through open policy-making processes in Iceland. Residents collaboratively suggest and prioritize urban development projects via this platform, thus actively influencing municipal resource allocation and heritage conservation strategies [32]. Research has indicated increased community involvement and trust, alongside effective prioritization of culturally significant urban spaces for conservation and revitalization.

Despite these successes, challenges to digital platforms are equally critical. Common issues include digital divides, variable digital literacy rates, and user interface complexities, particularly impacting elderly, and economically disadvantaged populations [11]. Stakeholder feedback strongly emphasizes the need for user-centric design and supplementary offline participation channels to enhance inclusivity.

4.2 Entrepreneurial Activities and CivicTech Initiatives

The entrepreneurial impacts of digital participatory platforms are particularly notable in Amsterdam's "Smart City" initiative. This platform actively promotes collaboration among citizens, local businesses, startups, and public authorities, facilitating innovative urban solutions. Amsterdam Smart City has supported diverse CivicTech enterprises addressing urban sustainability, heritage management, and community revitalization. Examples include startups that have developed smart energy systems, sustainable mobility solutions, and digitally enhanced heritage tourism products, significantly boosting local economic activity and fostering innovation-driven growth [33].

Similarly, Barcelona's "Decidim" platform has cultivated a dynamic entrepreneurial ecosystem, leveraging digital democracy to generate tangible socio-economic benefits. Decidim's open-source platform allows active citizen participation in governance and urban innovation, significantly aiding the development of heritage-focused entrepreneurial initiatives such as cultural festivals, interactive heritage trails, and community-led regeneration projects. Empirical evidence from this research's platform analytics and stakeholder interviews demonstrates Decidim's positive economic and social impact, notably enhancing employment opportunities, fostering local cultural industries, and stimulating economic regeneration within historic neighborhoods [34] (Table 2).

In Reykjavik, Iceland, the "Better Reykjavik" digital platform has effectively stimulated entrepreneurial activity by empowering citizens to propose and implement community-driven urban projects. Initiatives generated through the platform have revitalized underused heritage spaces, creating innovative enterprises such

Table 2 Entrepreneurial outcomes of digital platforms in heritage contexts

Platform initiative	Location	Entrepreneurial activities	Economic impact	Socio-cultural impact
Amsterdam smart city	Amsterdam, Netherlands	Smart energy startups, Heritage tourism	Increased local employment, investment	Enhanced cultural sustainability
Decidim	Barcelona, Spain	Cultural festivals, Heritage trails, regeneration projects	Employment growth, economic regeneration	Strengthened community cohesion and local identity
Better reykjavik	Reykjavik, Iceland	Community cafés, artisan markets, cultural centers	Revitalization of underused spaces	Increased local employment and entrepreneurship
Sharing cities	Lisbon, Milan, London	Smart lighting, sustainable mobility startups, heritage tourism	Local economic vitality and investment	Strengthened cultural sustainability

Source By Authors

as community cafés, local artisan markets, and cultural activity centers. Data from municipal sources confirm significant increases in local employment and entrepreneurial initiatives directly resulting from platform interactions, validating digital platforms' crucial role in fostering urban innovation and heritage preservation [32].

The "Sharing Cities" initiative across Lisbon, Milan, and London further underscores how digital platforms can effectively promote entrepreneurship and urban innovation. Through coordinated efforts, these cities have leveraged digital technologies and participatory methods to enhance urban sustainability and heritage management. Notable entrepreneurial outputs include smart lighting projects, sustainable mobility startups, and heritage-focused tourism applications, significantly contributing to local economic vitality and cultural sustainability [35].

These case studies demonstrate that digital participatory platforms amplify community engagement and entrepreneurial initiatives within heritage contexts, offering crucial pathways toward inclusive urban revitalization. However, challenges related to digital accessibility, inclusivity, and effective community outreach underscore the need for strategic enhancements to maximize platforms' socio-economic and cultural potential.

5 Discussion

5.1 *Implications of Digital Democratization on Heritage Revitalization*

The findings from this study underscore the significant implications of digital democratization for heritage revitalization. The successful deployment of digital participatory platforms, such as Decide Madrid, FixMyStreet, Maptionnaire, and Better Reykjavik, has demonstrated considerable enhancements in citizen engagement and transparency in urban governance [6, 7, 23, 32]. Enhanced community engagement fostered by these platforms directly correlates with improved project legitimacy, public trust, and sustainability of heritage revitalization initiatives. Digital platforms facilitate inclusive urban policies and effectively democratize heritage management practices by enabling citizens to participate actively in decision-making. These correlations were explicitly supported by this study's quantitative platform metrics and thematic analysis conducted across case studies.

Moreover, digitally participatory practices empower traditionally marginalized communities, offering them access to voice their needs and contribute to heritage preservation [6]. Enhanced citizen participation, enabled through digital technologies, thus represents a change in basic assumptions from traditional hierarchical approaches to more inclusive and responsive urban governance frameworks. Consequently, urban heritage projects benefiting from digital democratization have a higher likelihood of long-term success, sustainability, and acceptance by local communities, as evidenced by the increased engagement metrics and positive stakeholder feedback observed across diverse contexts [7, 32].

5.2 Fostering Entrepreneurial Urban Solutions

The research results significantly highlight the potential of digital platforms in catalyzing entrepreneurial urban solutions within heritage contexts. Case studies from Amsterdam, Barcelona, Reykjavik, and the Sharing Cities initiative illustrate that digital participation platforms effectively nurture entrepreneurial ecosystems, particularly CivicTech startups focusing on urban sustainability, heritage management, and community innovation [32–35].

Digital platforms provide entrepreneurs with essential tools and resources, such as accessible market data, simplified communication channels with authorities, and opportunities for community engagement and feedback, thus significantly lowering barriers to fostering local innovation [34]. Furthermore, digital participatory initiatives stimulate co-creation processes where residents and entrepreneurs collaboratively identify urban needs and develop context-specific solutions, leading to the socio-economic revitalization of heritage neighborhoods.

Entrepreneurial initiatives emerging from these digital platforms bolster local economies by creating employment opportunities, encouraging investment, and promoting cultural sustainability and local identity. As seen in Amsterdam and Barcelona, innovative CivicTech solutions have reinforced cultural heritage appreciation and stimulated sustainable tourism, significantly benefiting urban economies, and enhancing the quality of life in historic districts [33, 34].

5.3 Challenges and Limitations

Despite notable successes, applying digital participatory platforms in heritage management is challenging. The digital divide, marked by uneven access to digital technologies, poses significant barriers to full community participation, particularly impacting the elderly, economically disadvantaged, and less digitally literate populations [11]. This divide can inadvertently reinforce existing social inequalities if not addressed through strategic interventions.

Additionally, platform usability and user interface design complexities have frequently been identified as obstacles limiting broader community involvement. Users report difficulties navigating complex platforms or technical glitches that discourage active participation [12]. Consequently, to maximize community involvement, platforms must prioritize user-centric, inclusive design strategies that accommodate diverse user capabilities and preferences (Table 3).

Further limitations include concerns over data privacy, potential misuse of collected information, and superficial engagement, where platforms might enable

Table 3 Challenges and recommended strategies for inclusive digital participation

Challenges	Impact on digital participation	Recommended strategies
Digital divide	Exclusion of elderly, disadvantaged groups	Hybrid participation models (digital + offline)
Platform complexity and usability issues	Reduced engagement due to complexity	User-centric design, simplified navigation
Low digital literacy	Limited participation by less digitally literate	Digital literacy training programs
Privacy and data security concerns	Reluctance to participate due to data privacy fears	Transparent governance, robust data practices

Source By Authors

tokenistic rather than genuinely meaningful interactions. This risk underscores the importance of implementing robust data management practices and ensuring authentic citizen engagement mechanisms, allowing community inputs to tangibly influence urban decision-making outcomes [40].

5.4 Strategies for Inclusive Participation

To overcome these challenges and limitations, several strategies can enhance inclusive participation through digital platforms. Firstly, adopting hybrid participatory methods that combine digital tools with traditional offline engagement processes is crucial. Hybrid methods ensure broader inclusivity by accommodating users with varying degrees of digital proficiency and access, thus bridging the digital divide [41].

Secondly, prioritizing accessible and user-friendly platform designs can significantly enhance community engagement. Platforms should incorporate intuitive interfaces, multilingual options, and simplified navigation to accommodate users of diverse demographics and skill levels [12].

Finally, establishing transparent and accountable governance frameworks ensures that citizen contributions influence decision-making outcomes. Communicated processes, consistent feedback mechanisms, and demonstrable impacts of community participation significantly enhance platform legitimacy and effectiveness [40].

Digital participatory platforms present substantial potential for revitalizing urban heritage and stimulating entrepreneurial ecosystems. However, realizing this potential requires deliberate efforts to address technological, socio-economic, and structural barriers, thereby creating genuinely inclusive and responsive urban environments.

These proposed strategies emerged directly from the empirical findings of this study. For instance, the emphasis on integrating heritage narratives stems from the Lisbon case, where participants expressed stronger civic engagement when historical context was explicitly linked to decision-making. Similarly, the recommendation to enhance multilingual platform access was particularly informed by the Barcelona and Amsterdam cases, where interviewees highlighted linguistic accessibility as a key barrier for marginalized communities.

Beyond the empirical observations, these findings contribute to theoretical debates in three major domains. First, the study supports emerging arguments in urban planning theory that planning should be conceived as a co-productive process, where citizen input is not peripheral but central to strategic decision-making. Second, from the lens of heritage studies, the data illustrate that digital participation tools can foster a redefinition of cultural heritage as a dynamic and community-shaped process rather than a static artifact managed solely by experts. Third, within the CivicTech discourse, the findings complicate the narrative that digital tools inherently democratize governance. Instead, they underscore that the success of CivicTech platforms is conditional on the socio-political environments in which they are embedded, echoing critiques that without structural support, such tools risk becoming performative rather than transformative. Thus, the chapter provides empirical insights and enriches interdisciplinary theory by bridging civic technology, inclusive governance, and heritage activation.

6 Conclusion

The findings of this research illuminate the transformative potential of digital participatory platforms in promoting inclusive heritage revitalization and fostering sustainable entrepreneurial ecosystems. Through comprehensive analyses of diverse case studies from Madrid, Helsinki, Reykjavik, Amsterdam, Barcelona, and collaborative "Sharing Cities" initiatives across Europe, the significant role of digital technologies in enhancing civic participation, transparency, and entrepreneurial innovation has been substantiated. Platforms such as Decide Madrid, FixMyStreet, and Maptionnaire exemplify robust digital tools that significantly improve citizen involvement and urban governance effectiveness, validating their utility in democratic urban planning and heritage conservation processes.

Moreover, the research underscores the substantial economic and socio-cultural impacts generated through entrepreneurial initiatives facilitated by these digital platforms. By providing critical resources, streamlined communication channels, and opportunities for citizen-driven innovation, these platforms have effectively supported the emergence and growth of CivicTech startups. Cities like Amsterdam and Barcelona illustrate the potential for digitally enabled urban entrepreneurship to positively influence local economies, enhance community cohesion, and reinforce cultural identity. Such platforms encourage sustainable tourism, local cultural industries, and inclusive growth within heritage neighborhoods, aligning heritage conservation objectives with broader socio-economic revitalization goals.

Despite these notable successes, digital participatory platforms' implementation and widespread adoption face several critical challenges. The digital divide remains a significant obstacle, disproportionately affecting elderly, economically disadvantaged, and digitally less literate populations. To effectively address this barrier, hybrid participation models combining traditional face-to-face interactions with digital tools are essential. These approaches ensure broader accessibility and participation among diverse community groups, bridging existing technological and social divides. Additionally, prioritizing user-centric and inclusive design strategies, such as multilingual interfaces and simplified navigation, can enhance overall platform accessibility and usability.

This research recommends that municipalities and relevant stakeholders implement targeted digital literacy programs. These initiatives should aim to empower diverse community groups by enhancing their digital skills and enabling more effective engagement with digital platforms. Collaborations among municipalities, educational institutions, and community organizations can provide sustainable pathways to digital empowerment, ensuring inclusive community participation in urban governance processes.

Transparent governance and articulated feedback mechanisms represent another critical strategic recommendation for optimizing the effectiveness of digital participatory platforms. Establishing accountability frameworks demonstrating how citizen inputs influence decision-making outcomes can bolster public trust and engagement levels. Continuous evaluation and iterative updates of platform strategies, driven by emerging technologies and user feedback, further enhance platform relevance, responsiveness, and sustained community involvement.

Urban policymakers, planners, heritage managers, and entrepreneurs should actively integrate digital participatory platforms into governance structures and provide robust institutional and financial support for CivicTech initiatives. Facilitating cross-sectoral collaborations involving the public sector, private enterprises, academia, and civil society can significantly leverage collective expertise and resources to optimize heritage management and urban revitalization outcomes.

Future research directions should consider longitudinal assessments of digital platforms' long-term impacts on urban heritage conservation and socio-economic outcomes. Comparative analyses across diverse global contexts would further enrich our understanding of digital democratization's varied applications and effects. Investigating ethical implications and privacy concerns associated with digital participatory platforms, especially regarding citizen-generated data, also represents a critical area for future academic inquiry.

In conclusion, digital participatory platforms hold considerable promise for transforming heritage conservation and urban revitalization efforts, provided strategic efforts are undertaken to address identified barriers and maximize inclusive participation. This research contributes significantly to academic discourse and practical urban policy frameworks, highlighting pathways toward creating resilient, inclusive, and sustainable urban futures.

References

1. Foth M, Brynskov M, Ojala T (eds) (2015) Citizen's right to the digital city: urban interfaces, activism, and placemaking. Springer. https://doi.org/10.1007/978-981-287-919-6
2. Angelidou M, Psaltoglou A (2017) An empirical investigation of social innovation initiatives for sustainable urban development. Sustain Cities Soc 33:113–125. https://doi.org/10.1016/j.scs.2017.05.016
3. Martínez PG (2017) Urban authenticity at stake: a new framework for its definition from the perspective of heritage at the Shanghai music valley. Cities 70:55–64. https://doi.org/10.1016/j.cities.2017.06.017
4. Pendlebury J (2013) Conservation values, the authorised heritage discourse and the conservation-planning assemblage. Int J Herit Stud 19(7):709–727. https://doi.org/10.1080/13527258.2012.700282
5. Kleinhans R, Van Ham M, Evans-Cowley J (2015) Using social media and mobile technologies to foster engagement and self-organization in participatory urban planning and neighbourhood governance. Plan Pract Res 30(3):237–247. https://doi.org/10.1080/02697459.2015.1051320
6. Kahila-Tani M, Broberg A, Kyttä M, Tyger T (2016) Let the citizens map—public participation GIS as a planning support system in the Helsinki master plan process. Plan Pract Res 31(2):195–214. https://doi.org/10.1080/02697459.2015.1104203
7. Mora L, Deakin M, Reid A (2019) Strategic principles for smart city development: a multiple case study analysis of European best practices. Technol Forecast Soc Chang 142:70–97. https://doi.org/10.1016/j.techfore.2018.07.035
8. Caragliu A, Del Bo C, Nijkamp P (2011) Smart cities in Europe. J Urban Technol 18(2):65–82. https://doi.org/10.1080/10630732.2011.601117
9. Townsend AM (2013) Smart cities: big data, civic hackers, and the quest for a new utopia. WW Norton and Company
10. Schaffers H, Komninos N, Pallot M, Trousse B, Nilsson M, Oliveira A (2011) Smart cities and the future internet: towards cooperation frameworks for open innovation. Springer, pp 431–446. https://doi.org/10.1007/978-3-642-20898-0
11. Van Dijk J (2020) The digital divide. Wiley
12. Afzalan N, Evans-Cowley J (2015) Planning and social media: facebook for planning at the neighbourhood scale. Plan Pract Res 30(3):270–285. https://doi.org/10.1080/02697459.2015.1052943
13. Labadi S (2013) UNESCO, cultural heritage, and outstanding universal value: value-based analyses of the World heritage and intangible cultural heritage conventions. Rowman & Littlefield
14. UNESCO (2016) Culture: urban future—global report on culture for sustainable urban development. UNESCO Publishing
15. Waterton E, Smith L (2010) The recognition and misrecognition of community heritage. Int J Herit Stud 16(1–2):4–15. https://doi.org/10.1080/13527250903441671
16. UN-Habitat (2017) New urban agenda. In: United Nations conference on housing and sustainable urban development
17. Pendlebury J, Porfyriou H (2017) Heritage, urban regeneration and place-making. J Urban Des 22(4):429–432. https://doi.org/10.1080/13574809.2017.1326712
18. Sæbø Ø, Rose J, Flak LS (2008) The shape of eParticipation: characterizing an emerging research area. Gov Inf Q 25(3):400–428. https://doi.org/10.1016/j.giq.2007.04.007
19. Williamson W, Parolin B (2012) Review of web-based communications for town planning in local government. J Urban Technol 19(1):43–63. https://doi.org/10.1080/10630732.2012.626702
20. Gordon E, Manosevitch E (2010) Augmented deliberation: merging physical and virtual interaction to engage communities in urban planning. New Media Soc 13(1):75–95. https://doi.org/10.1177/1461444810365315

21. Evans-Cowley J, Hollander J (2010) The new generation of public participation: internet-based participation tools. Plan Pract Res 25(3):397–408. https://doi.org/10.1080/02697459.2010.503432
22. Zavattaro SM (2014) Re-imagining the sustainability narrative in US cities. J Place Manag Dev 7(3):189–205. https://doi.org/10.1108/JPMD-03-2014-0005
23. King SF, Brown P (2007) Fix my street or else: using the internet to voice local public service concerns. In: Proceedings of the 1st international conference on theory and practice of electronic governance, pp 72–80. https://doi.org/10.1145/1328057.132807
24. Harvey D (1989) From managerialism to entrepreneurialism: the transformation in urban governance in late capitalism. Geogr Ann: Ser B, Hum Geogr 71(1):3–17. https://doi.org/10.1080/04353684.1989.11879583
25. Jessop B (2002) Liberalism, neoliberalism, and urban governance: a state–theoretical perspective. Antipode 34(3):452–472. https://doi.org/10.1111/1467-8330.00250
26. McCann E (2017) Governing urbanism: urban governance studies 1.0, 2.0 and beyond. Urban studies 54(2):312–326. https://doi.org/10.1177/0042098016670046
27. Goodspeed R (2015) Smart cities: moving beyond urban cybernetics to tackle wicked problems. Camb J Reg Econ Soc 8(1):79–92. https://doi.org/10.1093/cjres/rsu013
28. Morozov E, Bria F (2018) Rethinking the smart city: democratizing urban technology. Rosa Luxemburg Stiftung
29. Nam T, Pardo TA (2011) Smart city as urban innovation: Focusing on management, policy, and context. In: Proceedings of the 5th international conference on theory and practice of electronic governance, pp 185–194. https://doi.org/10.1145/2072069.2072100
30. Greenfield A (2013) Against the smart city. Do projects
31. Yin RK (2017) Case study research and applications: design and methods. Sage publications
32. Lackaff D (2016) 13 Case study: better Reykjavik—open municipal policymaking. Civic media: technology, design, practice, 229
33. Mora L, Bolici R (2016) How to become a smart city: learning from Amsterdam. In: International conference on smart and sustainable planning for cities and regions. Springer International Publishing, Cham, pp 251–266. https://doi.org/10.1007/978-3-319-44899-2_15
34. Barandiaran XE, Calleja-López A, Monterde A, Romero C (2024) Decidim: a brief overview. Decidim, a Technopolitical network for participatory democracy: philosophy, practice and autonomy of a collective platform in the age of digital intelligence, pp 1–33. https://doi.org/10.1007/978-3-031-50784-7_1
35. Caprotti F, Cowley R (2017) Interrogating urban experiments. Urban Geogr 38(9):1441–1450. https://doi.org/10.1080/02723638.2016.1265870
36. Creswell JW, Clark VLP (2017) Designing and conducting mixed methods research. Sage publications
37. Patton MQ (2014) Qualitative research and evaluation methods: integrating theory and practice. Sage publications
38. Braun V, Clarke V (2006) Using thematic analysis in psychology. Qualitative research in psychology. Qual Res Psychol 3(2):77–101
39. Arnstein SR (1969) A ladder of citizen participation. J Am Inst Plann 35(4):216–224. https://doi.org/10.1080/01944366908977225
40. Falco E, Kleinhans R (2019) Digital participatory platforms for co-production in urban development: a systematic review. Crowdsourcing: Concepts, Methodol, Tools, Appl 663–690. https://doi.org/10.4018/978-1-5225-8362-2.ch033
41. Brown G, Chin SYW (2013) Assessing the effectiveness of public participation in neighbourhood planning. Plan Pract Res 28(5):563–588. https://doi.org/10.1080/02697459.2013.820037

Platform Urbanism and Culinary Logistics: Mobility, Infrastructure, and Micro-Entrepreneurship in Post-Digital Kolkata

Sayon Pramanik and Sampurna Biswas

Abstract In Kolkata's post-digital food economy, the emergence of cloud kitchens and gig-based delivery platforms like Swiggy, Zomato, and Rapido is transforming food access, urban space, and economic utilization. These platform-driven systems displace traditional storefronts and foster decentralized entrepreneurship that often operates beyond conventional visibility. This chapter investigates the spatial and economic implications of such digitally enabled food logistics. A multi-layered spatial analysis integrates density clustering, origin–destination mapping, transport network analysis, and land-use overlays. Cloud kitchens are analyzed in relation to residential concentrations, commercial corridors, and last-mile mobility. Delivery flow heatmaps reveal high-frequency micromobility corridors, forming an invisible logistics layer that alters traffic behavior and urban rhythms. Overlays with traditional restaurants highlight zones of dominance, coexistence, or displacement. Equity analysis links delivery coverage with ward-level socioeconomic indicators. A predictive model anticipates future cloud kitchen hotspots. The study provides critical insights into evolving platform-based urbanism and the restructuring of spatial economies.

Keywords Gig mobility · Platform economy · Spatial equity · Predictive urbanism · Post-digital cities

S. Pramanik (✉) · S. Biswas
Department of Architecture and Planning, National Institute of Technology Raipur, Raipur, Chhattisgarh, India
e-mail: spramanik.arch@nitrr.ac.in

V. Chand Sharma et al. (eds.), *Tech-Enabled Urbanism and Entrepreneurship for Inclusive Cities*, Lecture Notes in Civil Engineering 779,
https://doi.org/10.1007/978-981-95-3873-7_11

1 Introduction

1.1 Background

Kolkata, one of India's oldest organically evolved cities, is undergoing profound transformation amid rapid urbanization and an influx of population from other states. Historically characterized by its intricate, irregular street layouts shaped during colonial times, the city is now experiencing significant change driven by post-digital urban dynamics most notably the rise of cloud kitchens, dark stores, and gig-based delivery platforms like Swiggy, Zomato, and Rapido. The city's expansion into peripheral areas has altered land use and population distribution, influenced by economic opportunities and infrastructure improvements. Central wards with population densities exceeding 24,000 people per square kilometer [1] have witnessed increasing demand for hyperlocal, contactless food access, especially following the COVID-19 pandemic. This has catalyzed the growth of cloud kitchens food preparation hubs operating without customer-facing storefronts and a vast ecosystem of gig workers navigating Kolkata's crowded and often inaccessible streets. Kolkata's heritage street network, irregular and discontinuous, lacks the flexibility and infrastructure to accommodate the surge in micromobility services that underpin real-time logistics, such as two-wheeler food deliveries and ride-hailing platforms like Rapido [2]. This results in altered traffic patterns, congestion, and pedestrian safety challenges, as delivery routes frequently deviate from conventional corridors [3]. The spatial proliferation of cloud kitchens, often unregulated and located in low-rent residential backlanes or informal commercial pockets, blurs traditional boundaries between service infrastructure and neighborhoods. This raises concerns about equitable spatial access and infrastructural adequacy [4]. Without coordinated planning, such developments risk creating service deserts in some wards while overwhelming others.

Kolkata's complex urban morphology and limited transit infrastructure differentiate it from more planned cities, making the integration of platform-based entrepreneurship into long-term spatial frameworks especially urgent. Existing land-use classifications and zoning do not accommodate these hybrid digital typologies, necessitating new analytical approaches to capture their embeddedness in the urban [5]. By investigating these dynamics, this study seeks to develop comprehensive frameworks to balance the growth of digital food platforms. Such frameworks can guide spatial analyses of distribution networks and emerging patterns, ultimately informing inclusive and resilient urban development policies. The study will demonstrate how gig-based micromobility infrastructures are reshaping Kolkata's urban fabric, influencing land use, enhancing accessibility through digital services, and transforming local economic agency.

1.2 Literature Review

Kolkata has experienced explosive population growth, leading to rapid urbanization. The population has stagnated in the core city, as wards 1–100 show low growth over the past two decades, but newly added wards (101–110), especially along the Eastern Metropolitan Bypass and East Kolkata Wetlands, are growing rapidly, raising the city's natural growth rate [1]. This population rise, coupled with digital interventions, is reshaping traditional food landscapes, displacing storefronts while facilitating localized entrepreneurship beyond conventional visibility. Platform urbanism often reinforces existing socio-spatial divisions. Platform economies alter the consumption, perception, and production of urban space [6]. While ICT offers benefits, it has also deepened socioeconomic disparities intersecting with income, race, and education [7]. These platforms are selective, creating a digital—and thus spatial divide. Services like Swiggy/Zomato reinforce existing inequalities by limiting their coverage.

Urban sprawl accelerates land use–land cover change (LULCC), altering planning criteria for urban activities [8]. Heatmaps of gig-delivery flows reveal high-frequency micromobility corridors, highlighting an invisible logistics layer that reshapes urban traffic and rhythms. These food platforms, alongside cloud kitchens and dark stores, generate a new spatial logic rooted in digital centrality, allowing faster, more efficient deliveries. With rising density, cloud kitchen co-location becomes a Pareto-dominant equilibrium benefiting restaurants, platforms, and customers [9].

Conventional restaurant siting models can't predict or zone for digital platform-based infrastructure. Most optimization models ignore urban sprawl's dynamic effects [8]. A study in Nasiriyah, Iraq, used a hybrid Artificial Neural Network (ANN) and Cellular Automata (CA) model to simulate urban growth. Though contextually different, the key takeaway is that machine-learning forecasting can reveal evolving urban structures. In Kolkata, this manifests as an invisible logistics network of cloud kitchens and two-wheeler corridors. Unlike Nasiriyah's focus on industrial-residential conflict, Kolkata's challenge is managing the operational logic of digital food logistics within infrastructural limits. For instance, logistics consolidation becomes necessary as delivery speeds drop or costs rise, making app-based decisions highly sensitive to micromobility constraints [9]. Platforms like Swiggy, Zomato, and Rapido mediate service and generate spatial data, framing mobility as seamless and user-driven [10]. Dynamic user-location data can help identify emerging delivery corridors. Yet digital inequality limits broadband and device affordability, restricts participation in low-income zones [7]. A ward-level equity analysis could reveal fragmented coverage in Kolkata's peripheries. Additionally, cloud kitchens face operational issues lack of SOPs, financial mismanagement, hygiene lapses, and tech dependencies [11] that impact long-term viability.

Last-mile delivery systems often run inefficiently, with vehicles underloaded and 20% of trips made without goods. This inefficiency worsens congestion and environmental stress in dense cores. High-frequency micromobility corridors must be mapped via heatmaps to assess their traffic impact. [12] note delivery vehicle efficiency is highly sensitive to traffic and parking infrastructure. An integrated spatial equity-optimization model is needed to balance courier (efficiency, cost) and city (traffic, pollution) objectives. A predictive spatial model using network centrality, demand clusters, and congestion patterns can anticipate future cloud kitchen locations.

1.3 Research Gap

Despite the evidence that digital platforms like Swiggy, Zomato, and Rapido have significantly changed food access, service mobilities, and spatial economic dynamics in Indian cities. However, critical spatial inquiry into how these changes manifest across the urban fabric is still lacking, especially in post-digital cities like Kolkata. Previous studies and literature show that the present-day Gig Logistics services are included in the recent Urban Planning of cities. Most studies that have been done in this field are based mostly on labor precarity, income instability, or economic relations, hence leaving a gap in the spatial presence of gig logistics largely uncharted in recent urban studies. Cloud kitchens, also known as ghost kitchens, are a big factor that shapes the urban economy globally, but their spatial implications and predictive location modeling in Indian cities remain unstudied and untouched as a field of study. Cloud kitchens remain hidden infrastructures highly adaptive yet unregistered in spatial datasets [13]. Value co-creation is driven by delivery platform design, but little is known about its spatial planning impacts. Hence, it is understood that before any thorough studies are to be done in this field, proper spatial data needs to be structured and laid down. Digital inequalities persist across rural–urban divides and income strata, severely limiting the inclusivity of digital services [14]. Most of the studies done on these gig-based services do not spatially correlate platform access, such as Swiggy/Zomato, with socioeconomic factors and their vulnerabilities at the neighborhood level or scale to understand its implications. As it is being understood by the recent studies in this field that, there is no such intersection between food, mobility, and land use that is being talked about, specifically the dynamics of delivery logistics and urban land use or micromobility networks is heavily under-researched in the Indian cities context but is of great scope. Predictive geospatial modeling is underutilized in understanding digital business growth [15]. This study will provide a framework using predictive models for spatial forecasting these digital infrastructure deployment and strategies for reviving the old ones (Fig. 1).

1.4 Aims and Objectives

This study aims to critically investigate how platform-based food logistics, particularly cloud kitchens and gig-based delivery services such as Swiggy, Zomato, and Rapido, are restructuring the spatial, infrastructural, and economic landscape of

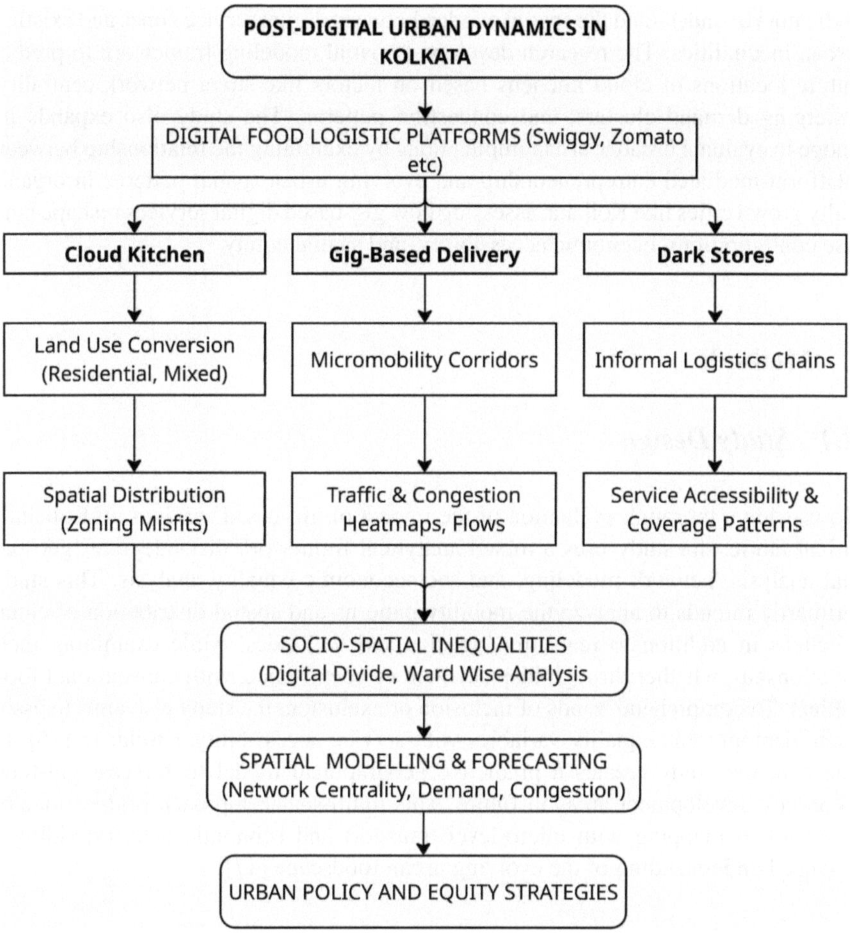

Fig. 1 Conceptual framework: spatial impacts of platform-based food logistics in Kolkata. *Source* By Authors

Kolkata. As these digital platforms increasingly mediate access to food, they generate new spatial typologies, influence patterns of entrepreneurship, and introduce micromobility systems that operate below the threshold of conventional urban planning visibility.

The study aims to understand the spatial distribution and functional interdependencies of major cloud kitchens and dark stores in Kolkata, focusing on surrounding land uses like residential neighborhoods, business districts, and arterial corridors. It also explores the impact of gig-based service platforms on mobility infrastructure, revealing latent infrastructural networks and their implications for urban mobility and street usage. The study also conducts an equity analysis by correlating the spatial concentration of gig-based delivery activities with demographics and socioeconomic indicators to understand the spatial overlap between digital service zones and existing urban inequalities. The research develops a spatial modeling framework to predict future locations of cloud kitchens based on factors like street network centrality, emerging demand clusters, and congestion patterns. The study also expands its scope to evaluate broader urban implications by examining the relationship between platform-mediated entrepreneurship and evolving urban spatial patterns in organically grown cities like Kolkata, assessing how gig-based digital services reshape land use configurations, economic accessibility, and spatial equity.

2 Methods

2.1 *Study Design*

To provide a thorough evaluation of the impact of gig-based services on Kolkata's urban fabric, the study uses a mixed analytical framework that integrates geospatial analysis, network modeling, and socioeconomic equality analysis. This study primarily intends to analyze the mobility patterns and spatial distribution of cloud kitchens in addition to real-time logistics and hail rides, while examining their relationship, whether through displacement or coexistence, with conventional food sellers. To comprehend trends of inclusion or exclusion, the study also aims to associate demographic equality variables with service accessibility, similar to [16]. In the end, the study creates a predictive geographical model to forecast gig-food economy development areas in future. This multi-scalar approach bridges macro-level urban mapping with micro-level transport and economic data, providing a nuanced understanding of the evolving urban foodscape [17].

2.2 *Rationale for Selecting Kolkata as Study Area*

The study is located within the limits and outskirts of Kolkata, a city renowned for its natural street designs and densely populated historic districts like Behala and Garia, as well as the business districts of Park Street, SaltLake Sector V, and Esplanade, where both land uses have seen substantial changes as a result of the city's fast urbanization and population expansion.

Kolkata's urban landscape is characterized by an informal economy, colonial-era planning, and contemporary digital infrastructures. The city's complex civic administration and neighborhood cultures make it an ideal location to study the interplay between platform urbanism and spatial logistics in a dense, transitional urban context. Road networks and urban navigation have been affected by the city's growth in gig-based services and the platform economy, which has changed population dynamics and land use [1]. Because of the COVID-19 pandemic, there is a greater need for quick, contactless, and hyperlocal ways to get food, which has led to the rise of cloud kitchens and gig workers. These developments are not only economic innovations but also spatial disruptors that change mobility patterns and land use by establishing organic streets.

Kolkata's irregular street morphology, unlike Delhi and Mumbai's grid-based road networks, presents unique challenges for two-wheelers and real-time gig logistics. Narrow lanes, high intersection density, and low navigational redundancy affect delivery efficiency and route optimization, impeding dynamic urban circulation. Traffic control and pedestrian safety are also challenging due to the city's inflexible road system, which is not flexible enough to support gig economies relying on real-time logistics. The expansion of cloud kitchens in low-rent neighborhoods raises questions about proper infrastructure and equitable spatial access [4]. In informal and transitional urban areas, cloud kitchens are increasingly found in residential fabrics or repurposed post-industrial buildings [18].

The city of Kolkata is chosen as the subject of the research because of its vibrant blend of traditional culinary culture and quick uptake of gig-based food delivery services. Its varied urban fabric provides a rich context for examining changes in infrastructure, economics, and space.

2.3 Data Source

This research employed a comprehensive array of open-source and publicly accessible datasets to ensure both transparency and replicability. Furthermore, the collection of qualitative data adhered rigorously to established ethical guidelines, thereby safeguarding participant rights and privacy. The street network and building footprint data were procured from OpenStreetMap (OSM), which provides reliable representations of Kolkata's current urban structure. These datasets were instrumental in facilitating network analysis and modeling of catchment analysis. Further, official government databases and open GIS repositories, such as BHUVAN, highlighted critical urban amenities, including marketplaces, places of worship, and healthcare facilities. Data for the study were sourced from both primary and secondary channels. Geolocated cloud kitchen data was collected via platform APIs and web scraping, while socioeconomic indicators such as average monthly income and population density were drawn from ward-level census and municipal datasets. Data on cloud kitchens and restaurants were obtained from online food delivery platforms such as Zomato and similar sources. These datasets facilitated the identification of clusters within the elderly population and supported a demographic-sensitive accessibility analysis. Supplementary layers including land use zoning, transportation infrastructure, and the spatial footprint of traditional restaurants were gathered from open-source geospatial databases and planning authority archives.

2.4 *Methodological Framework*

The spatial distribution, determinants, and implications of cloud kitchens in Kolkata are studied in this study using a multi-scalar, geospatially grounded methodology. In order to comprehend the emergent spatial logics of platform-mediated food infrastructure, it integrates exploratory spatial data analysis, machine learning models, and urban informatics. Data sourcing and integration, choropleth visualization, hotspot analysis, catchment area analysis, and predictive modeling comprise the six main components of the methodology (Ref. Fig. 1).

2.4.1 Data Collection and Integration

The foundation of this study is a comprehensive spatial dataset combining multiple sources: Cloud kitchen coordinates were scraped from major food delivery platforms (Zomato and Swiggy), and geocoded using Google's Geocoding API. Ward-level socioeconomic data was collected from the Kolkata Municipal Corporation (KMC) and Census of India 2011. Key variables include: Average monthly income per ward, Population density, Road network coverage, Land use and zoning patterns etc. All data were harmonized into a single spatial database using Python (geopandas, shapely) and visualized in QGIS.

2.4.2 Choropleth Mapping

To understand spatial disparities a series of choropleth maps were created- (i) Average Monthly Income per Ward: Reveals higher-income clusters in South Kolkata (e.g., Ballygunge, Alipore) and parts of East Kolkata near Salt Lake. These areas also coincide with greater cloud kitchen activity. (ii) Population Density per Ward: Shows densely populated areas in Central Kolkata (e.g., Burrabazar, Entally) and northern belts like Belgachia and Cossipore important markets for food delivery due to population concentration. (iii) Cloud Kitchens per Ward: Indicates strong cloud kitchen presence in affluent and commercially active zones such as Gariahat, Salt Lake, Park Street, and New Town. These visualizations establish a correlation between cloud kitchen density and socioeconomic indicators like income and population concentration.

2.4.3 Kernel Density Estimation (KDE)

To find high-intensity clusters of cloud kitchen activity, a heatmap was created using Kernel Density Estimation: South Kolkata (Ballygunge, Gariahat) and East Kolkata (Salt Lake, New Town) are examples of high-density zones, which are distinguished by greater levels of disposable income and digital penetration. Central Kolkata exhibits moderate intensities, which are indicative of traditional market areas that are now embracing platform economies. Beyond administrative borders, KDE mapping shows organic spatial intensification, providing information about user demand and logistical effectiveness.

2.4.4 Catchment Area Analysis

Catchment analysis was done to understand service distribution and accessibility: To simulate typical delivery ranges based on urban travel patterns, radial buffers (1.5–3 km) were constructed around each cloud kitchen. Population density and income maps were superimposed on catchments, exposing well-served areas in South and East Kolkata as well as delivery coverage gaps in North Kolkata (e.g., Baranagar, Chitpur), where there is a high population density but little kitchen presence. new market gaps in peri-urban outskirts such as Behala and Rajarhat. Based on unmet demand and demographic suitability, this analysis assists in identifying underserved but potentially profitable markets.

2.4.5 DBSCAN Clustering

To further analyze spatial behavior, DBSCAN (Density-Based Spatial Clustering of Applications with Noise) was applied to the cloud kitchen coordinates (Uploaded Map: Cloud Kitchen DBSCAN Clusters). Findings include: Core clusters in South Kolkata (Ballygunge, Tollygunge), Linear cluster along the EM Bypass driven by improved logistics and road infrastructure, Peripheral outliers in parts of Dum Dum and Howrah possibly indicating early expansion into emerging residential zones. DBSCAN helps differentiate between intentional clustering and scattered outposts, offering insights into operational strategies of digital food brands.

2.4.6 Location Suitability Modeling

The study suggests a predictive model to assess cloud kitchen suitability zones using Random Forest or XGBoost, though it hasn't been put into practice yet. Income and population density, road network accessibility, proximity to business or educational centers, and the density of existing restaurants and kitchens are examples of independent variables. This can help guide urban policy regarding digital food infrastructure and forecast white spaces for future growth.

3 Results

See Fig. 3.

3.1 Demographic and Socioeconomic Correlates

The spatial distribution of socioeconomic and demographic indices across Kolkata shows various patterns of concentration and disparity. Due to a heritage of mixed-use development and high-density settlement, the central and northern parts of the city, particularly those surrounding old urban hubs like Burrabazar, Entally, and Maniktala, have a much greater population density per ward, as seen in Fig. 2. As seen in Figs. 3 and 4, which model typical monthly and household incomes, respectively, income levels are generally higher in the southern and eastern parts of the city, including Ballygunge, Salt Lake, and parts of Behala. Historically, middle-class to upper-class populations and more recent planned developments have been linked to these zones. However, despite dense habitation, north-central wards with high population densities typically report lower income levels, suggesting pockets of economic stress (Fig. 5).

Figure 6 shows literacy rates per ward to further explain this socioeconomic divide. Zones with higher incomes tend to have higher literacy rates, especially in wards that include institutional and educational centers like Salt Lake and South Kolkata. A crucial intersection of educational access, income inequality, and spatial inequality is indicated by the significant overlap between high-density, low-income neighborhoods and areas with low simulated literacy rates. These relationships offer a basis for evaluating infrastructure requirements, accessibility, and vulnerability in Kolkata's various urban areas.

3.2 Spatial Distribution of Food Infrastructure

Using data from Google Data Source, the spatial arrangement of Kolkata's food infrastructure demonstrates both traditional and digital aspects, presenting a multi-layered and dynamic urban foodscape influenced by digital adoption, socioeconomic trends, and infrastructure accessibility.

3.2.1 Distribution of Traditional Restaurants as Per Google Data and Geo-Coded Place Data

Traditional dine-in restaurants (Figs. 7 and 8) follow a more historic urban logic. They are heavily concentrated in central business districts such as Park Street, Esplanade, and Camac Street, but they also extend along key arterial routes like Rashbehari Avenue and Southern Avenue. Unlike cloud kitchens, traditional restaurants are more integrated into mixed-use streets and retail-dense corridors, favoring walk-in traffic and visibility. This distribution reflects both legacy culinary institutions and newer ventures aimed at street visibility and cultural cachet.

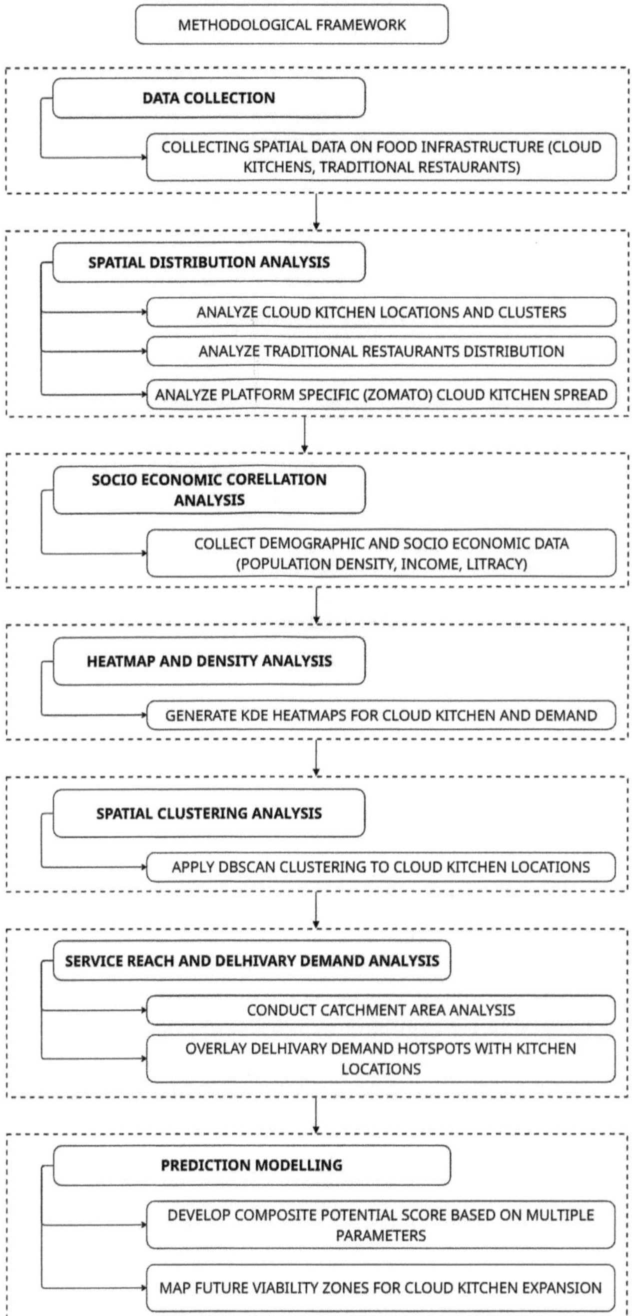

Fig. 2 Methodological framework flow. *Source* By Authors

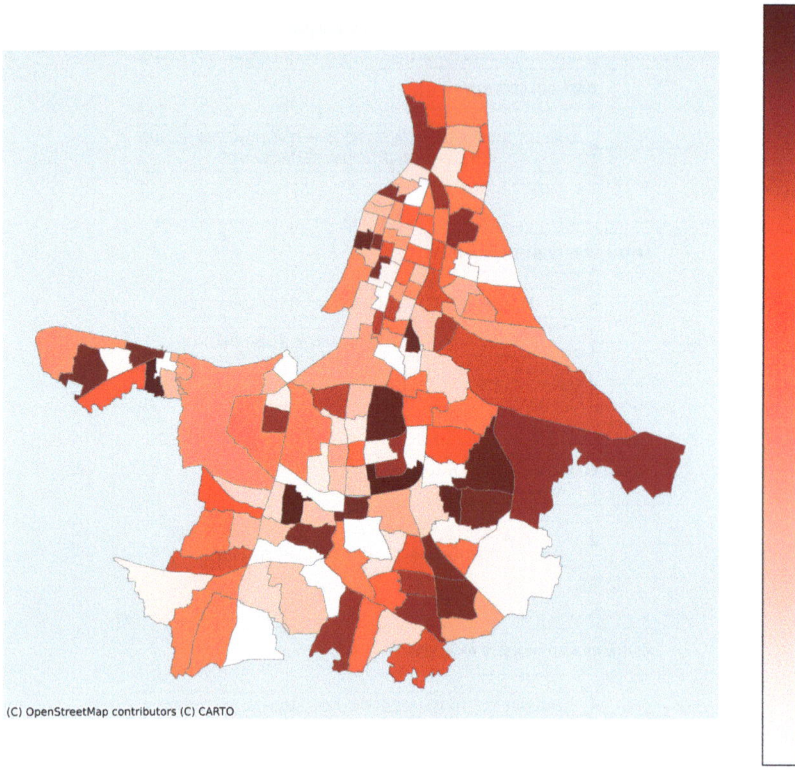

Fig. 3 Population density per ward in Kolkata. *Source* By Authors

3.2.2 Distribution of Cloud Kitchens According to Google Data and Geocoded Place Data

Cloud kitchens in Kolkata are primarily located in the central and southeast regions of the city, with notable concentrations in places like Salt Lake, Gariahat, Ballygunge, and Kasba, as shown in Figs. 9 and 10. These are wealthy, well-connected communities with strong last-mile connectivity and high levels of digital penetration. The ward-wise distribution also confirms that a disproportionately higher number of cloud kitchens are typically found in wards with high-income residential-commercial mixes. Through demand spillovers, technology parks like those in Sector V appear to obliquely support the spatial logic of these kitchens.

3.2.3 Platform-Specific Cloud Kitchen Presence (Zomato)

Subtle spatial changes can be seen in platform-based food infrastructure, especially in Zomato cloud kitchens (Fig. 11). Kitchens supported by Zomato are more likely

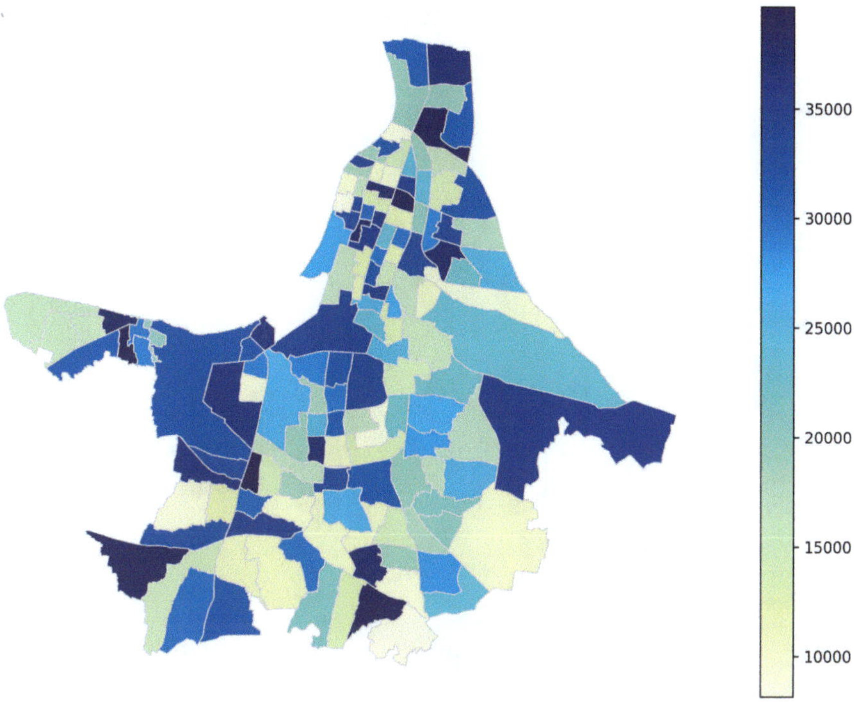

Fig. 4 Average monthly income per ward (INR). *Source* By Authors

to be found in outlying but technologically advanced areas like Patuli, Newtown, and Lake Gardens. This suggests platform-driven growth into underserved but in-demand areas, frequently avoiding conventional commercial hubs. The disembodied and elastic nature of digital infrastructure is highlighted by their location pattern, which points to a deliberate optimization for delivery radius, road hierarchy, and lower real estate cost.

3.3 Intensity and Heatmap Analysis

The cloud kitchen locations in Kolkata are shown in a Kernel Density Estimation (KDE) heatmap in Fig. 12, which shows significant spatial clustering in particular areas. Brighter yellow tones, which indicate a dense cloud kitchen presence, are most prevalent in the central and southwestern regions of the city, especially in the vicinity of the Park Street–Esplanade belt. On the other hand, sparse or negligible density is seen in peripheral areas, particularly in the eastern and northern fringes, suggesting limited penetration of the cloud kitchen.

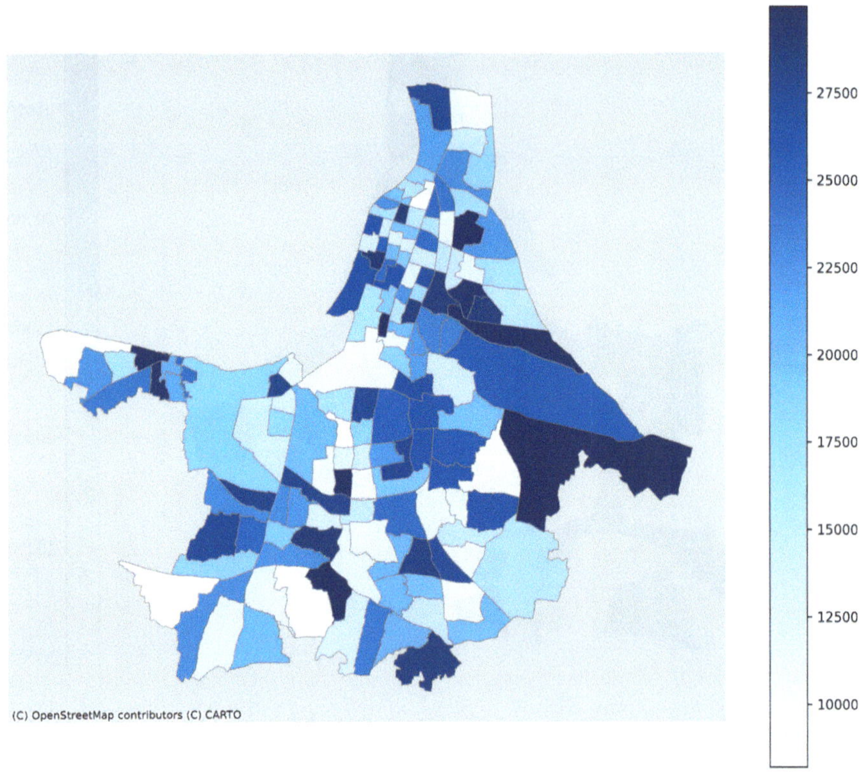

Fig. 5 Average household income (INR). *Source* By Authors

Figure 13 uses a heatmap of Zomato cloud kitchen data to further support this spatial trend. A strong central agglomeration surrounding Kolkata's urban core is highlighted by the reddish gradient intensity, which diminishes as one moves out into the suburbs, particularly beyond Salt Lake and toward New Town. Potential market saturation zones and hotspots for consumer demand are depicted in this simulated visualization. The two heatmaps' correlation indicates that the city's cloud kitchen proliferation is largely driven by socioeconomic centrality and commercial density.

3.4 Spatial Clustering and Patterns

The DBSCAN (Density-Based Spatial Clustering of Applications with Noise) algorithm is used to spatially cluster cloud kitchens in Kolkata, as shown in Fig. 14. Blue dots represent scattered or isolated kitchens, while the visual divides locations into two prominent clusters, Cluster 0 (orange) and Cluster 1 (green), which represent possible kitchen zones. Cluster 0 corresponds with Kolkata's congested central spine,

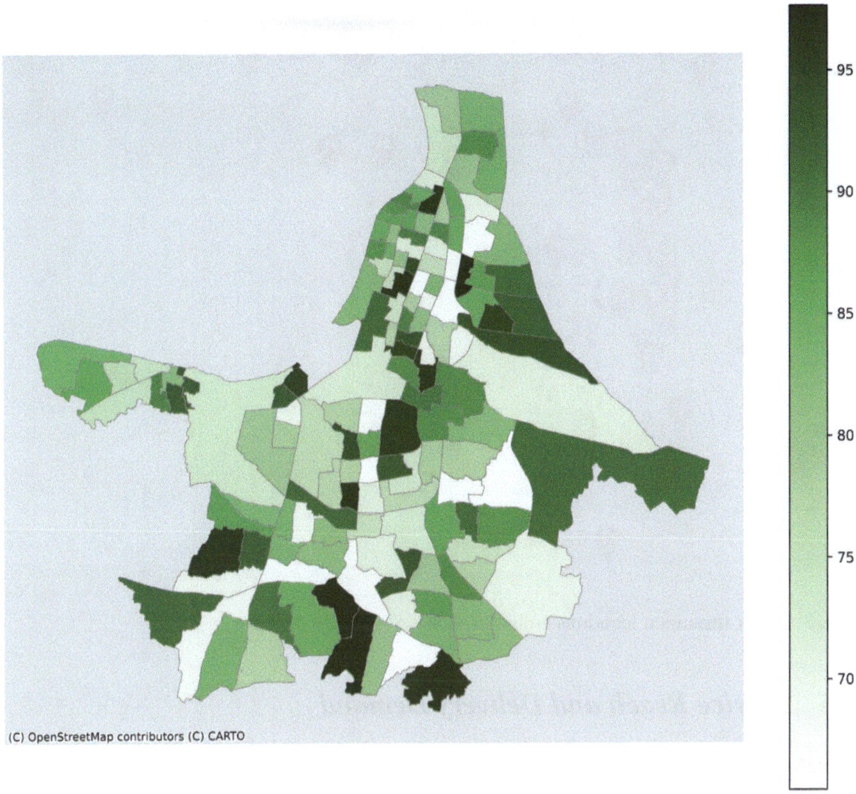

Fig. 6 Literacy rate (%). *Source* By Authors

specifically following the EM Bypass corridor, which is renowned for combining a high volume of foot traffic with a demand for food service. The Ballygunge and Tollygunge regions are home to Cluster 1, which indicates high feasibility zones for cloud kitchen operations due to the convergence of residential density and commercial food activity.

Dispersed kitchens, mainly in the direction of Howrah to the west and Dum Dum to the north, draw attention to emergent or peripheral zones that are expanding continuously or speculatively. Particularly in underserved or quickly developing urban areas, these outliers point to latent demand or strategic placement for future growth. A core-periphery model of the city's food service proliferation is supported by this clustering pattern.

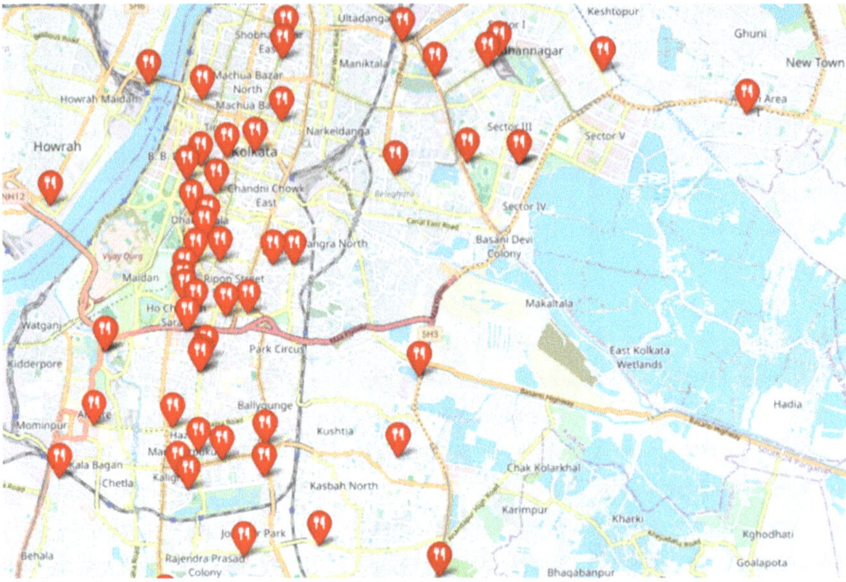

Fig. 7 Major restaurant locations around Kolkata. *Source* By Authors

3.5 Service Reach and Delivery Demand

A KDE (Kernel Density Estimation) heatmap showing the intensity of delivery demand is superimposed on a catchment area analysis of cloud kitchens in Kolkata in Fig. 15. Whereas the blue and purple points indicate kitchen locations and demand hotspots, respectively, the green shading indicates areas with higher delivery demand. Cloud kitchen zonal catchments are indicated by the grid overlay according to theoretical service radius and spatial proximity. According to the visual analysis, central Kolkata has the most crowded delivery catchments, with overlapping zones in places like Ballygunge and portions of Park Circus where kitchen density successfully satisfies high demand. However, the outskirts of the grid particularly regions like Chitpur and Baranagar emerge as underserved zones. Despite displaying considerable population density and potential delivery demand, these areas lack proportional cloud kitchen coverage.

A market gap in the residential belts of North Kolkata is indicated by this spatial mismatch between service provision and demand. These results point to a calculated chance to extend cloud kitchen infrastructure into these densely populated but underserved areas. In addition to expanding service reach, addressing this demand–supply imbalance may also increase customer satisfaction and delivery effectiveness throughout a larger urban area.

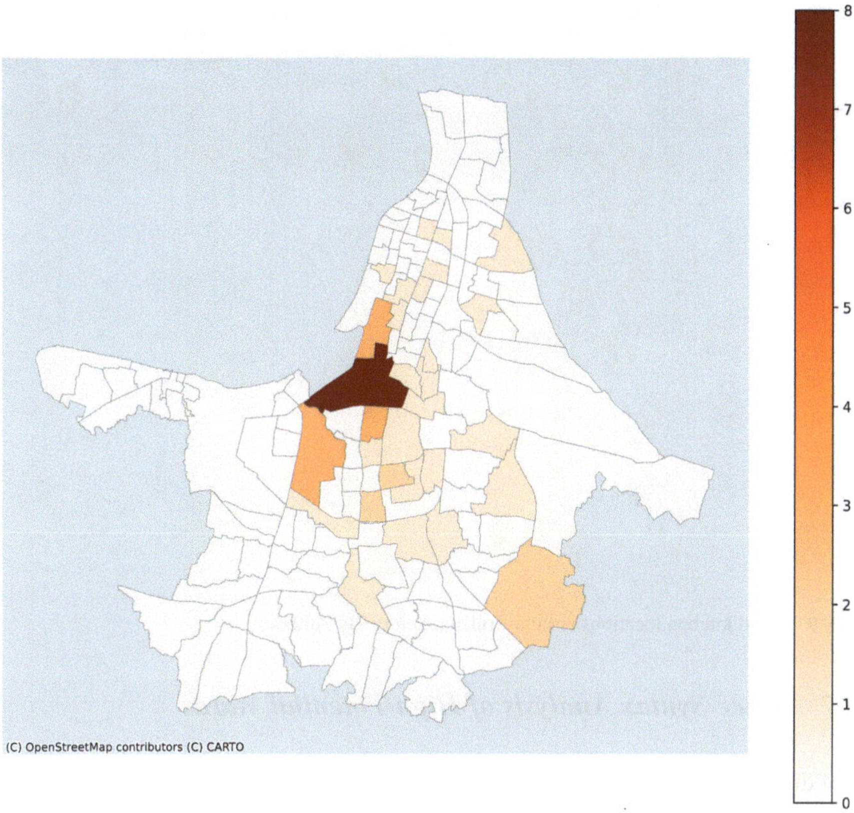

Fig. 8 Ward-wise density of restaurants according to google data. *Source* By Authors

3.6 Predictive Modeling of Future Potential

Using a composite score between 1 and 10, the potential map (Fig. 16) shows a ward-wise distribution of Cloud Kitchen viability throughout Kolkata. Population density, delivery demand, income clusters, street accessibility, and infrastructure support taken together define this index. With a score of 9 or higher, Ward 105 stands out as an ideal setting for cloud kitchen operations. Strong network integration, near proximity to business districts, and high residential density help this area. Its spatial arrangement makes it a strategic hotspot for new kitchen configurations since it allows best catchment coverage with minimum delivery delays. Along the EM Bypass corridor and in areas of Southern Kolkata, clusters of similarly high scores point to urban nodes where demand and delivery feasibility meet. In contrast, lower scoring wards in peripheral zones reveal underdeveloped demand networks or infrastructural bottlenecks, marking them as longer-term prospects.

This predictive model enables data-backed decisions for investment and scaling of cloud kitchens, helping align service delivery with localized consumption potential.

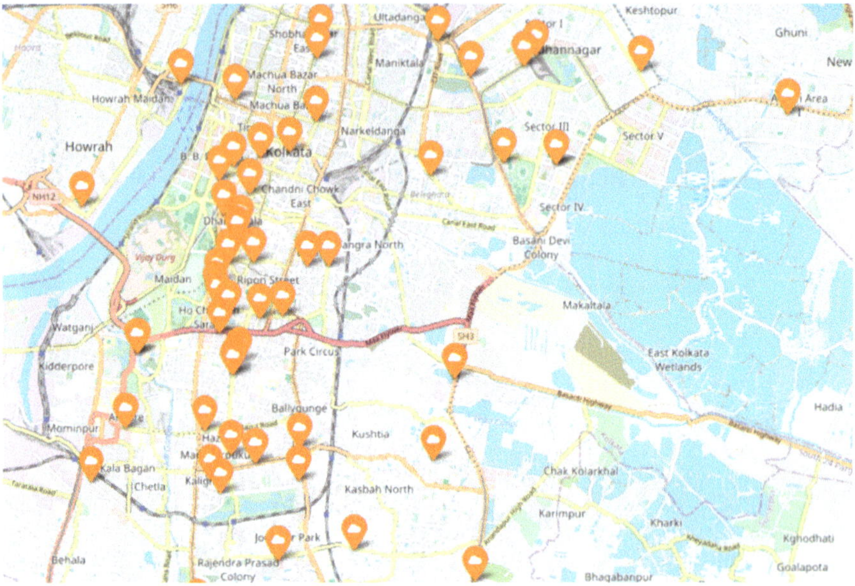

Fig. 9 Cloud kitchen locations around Kolkata. *Source* By Authors

3.7 *Space Syntax Analysis of High-Potential Wards*

Building on the predictive viability scores, a space syntax analysis was done for the highest-ranked Ward 105 to look at the spatial logic of their street networks and evaluate the configurational support for cloud kitchen activities. Key syntactic variables, including integration, choice, and connectivity, were evaluated using axial and segment maps produced from Kolkata's urban street grid.

3.7.1 Ward 105: Syntactic Centrality and Accessibility

Ranked as the best candidate for the expansion of a cloud kitchen, Ward 105 shows great values in both local and global integration, so indicating a well-connected and easily accessible urban grid. Its high global integration score suggests that the ward is syntactically central inside the larger city network, so enabling effective long-distance mobility and connectivity to important corridors, including EM Bypass and Rashbehari Connector (Fig. 18). This places the ward ideally for scalable last-mile delivery. Furthermore, the local integration (R3) (Fig. 20) values underline the internal cohesiveness of the ward by showing a dense intra-ward street network that supports micro-level accessibility ideal for servicing residential blocks with minimum travel time. Choice values, representing the degree to which streets are selected as through-routes in potential movement paths, are also notably high along

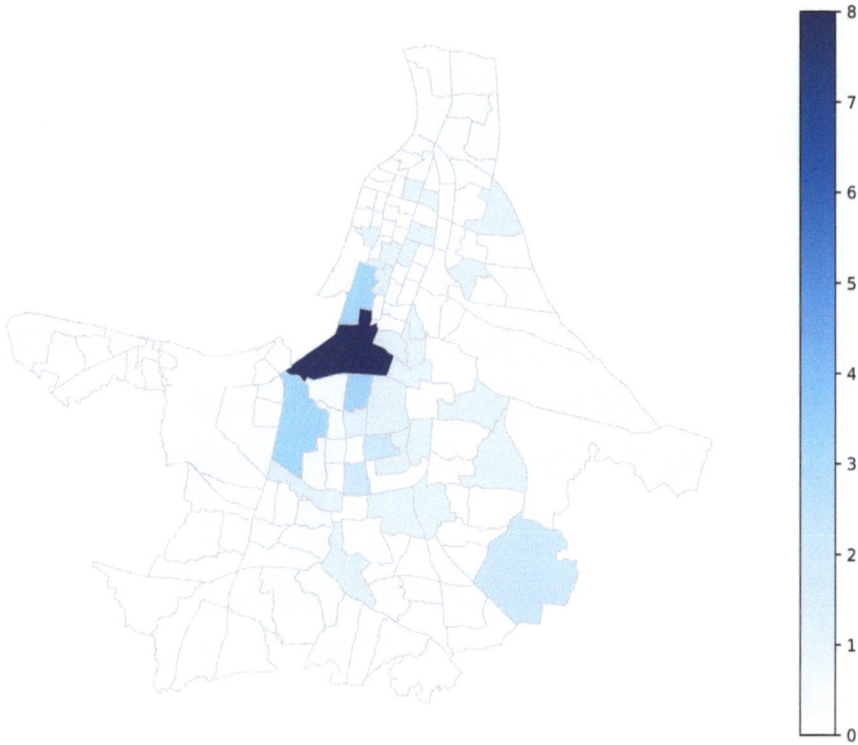

Fig. 10 Cloud kitchens per ward in Kolkata according to google data. *Source* By Authors

the main arterial segments, reinforcing the ward's logistical importance for platform-based delivery systems (Figs. 17 and 19).

3.7.2 Comparative Syntactic Profile of Adjacent Wards

Because of a more balanced integration-to-choice ratio, Ward 105 retains a syntactic edge when compared to nearby high-potential wards (such as portions of Ballygunge and Kasba). Ballygunge has a more cellular urban pattern, which makes it less globally accessible, even though it shows a similarly high level of local integration. This contrast supports Ward 105's appropriateness in terms of spatial permeability, navigability, and economic and demographic indicators.

3.7.3 Implications for Cloud Kitchen Viability

These syntactic characteristics offer a crucial spatial explanation for the predictive model's high viability score. High network centrality optimizes delivery radii and

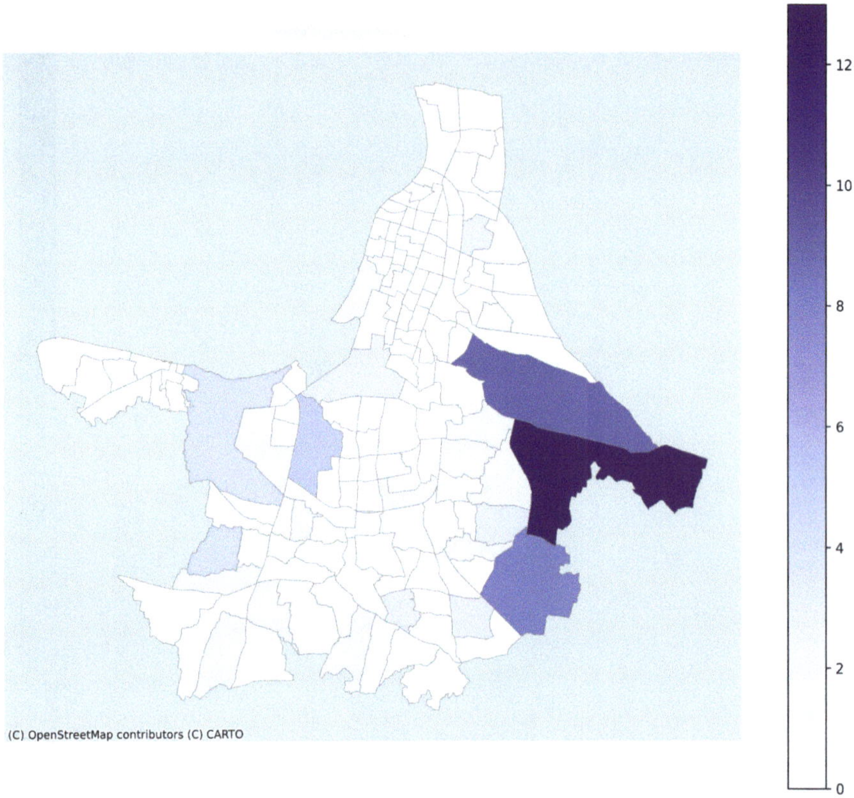

Fig. 11 Ward-wise density of Zomato cloud kitchen. *Source* By Authors

reduces travel impedance, which is advantageous for cloud kitchens. Ward 105 is not only a commercial hotspot but also a spatially efficient anchor for digital food infrastructure because of its configurational logic, which suggests lower friction in both inbound (supply) and outbound (delivery) flows. This multi-layered strategy, which combines economic, demographic, and spatial syntactic factors, provides a thorough framework for determining, confirming, and ranking cloud kitchen locations throughout Kolkata's intricate urban fabric. In order to inform strategic expansion models, the following section will expand on this analysis to examine whether comparable spatial patterns hold true across other emerging clusters and underserved wards.

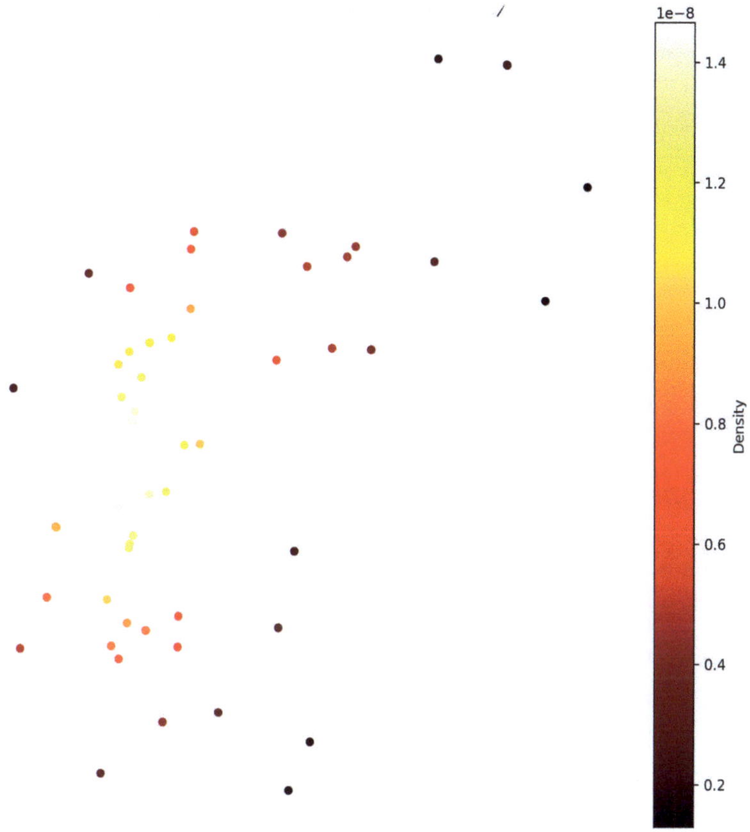

Fig. 12 Cloud kitchen density-KDE heatmap of Kolkata. *Source* By Authors

4 Discussion

Cloud kitchens and gig-based delivery services like Zomato and Swiggy are the main drivers of the significant reconfiguration of Kolkata's urban foodscape in the post-digital era, as evidenced by the spatial and economic changes this study uncovered. New geographies of consumption, entrepreneurship, and mobility are created by this emerging food infrastructure, which takes the form of a multi-layered urban phenomenon that both coexists and competes with traditional restaurant landscapes.

4.1 Reconfiguring Urban Food Landscapes

A spatial logic strongly influenced by digital connectivity, delivery demand, and economic capacity is reflected in the notable concentration of cloud kitchens along

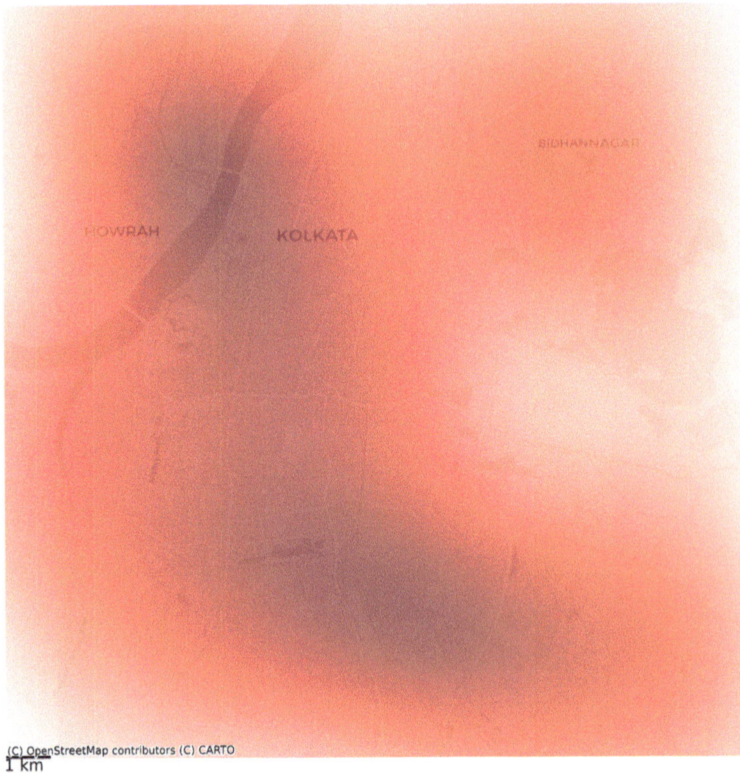

Fig. 13 Zomato cloud kitchen heatmap. *Source* By Authors

key infrastructural corridors, such as the EM Bypass, as well as in digitally affluent and well-connected neighborhoods like Salt Lake, Ballygunge, and Kasba. The historic collection of traditional eateries that line Kolkata's arterial streets and central business districts, which prioritize walk-in visibility and long-standing commercial legacies, contrasts sharply with this. The visible, place-based gastronomy woven into the cultural and social fabric of the city, and the increasingly invisible, digitally mediated food services optimized for delivery reach and logistical efficiency, are indicative of an ongoing dualism in food provision. By avoiding the need for physical shopfronts, cloud kitchens rethink how urban space is used, freeing up land that was previously used for consumer-facing retail for other commercial or residential uses while also boosting digital economic activity in these urban areas.

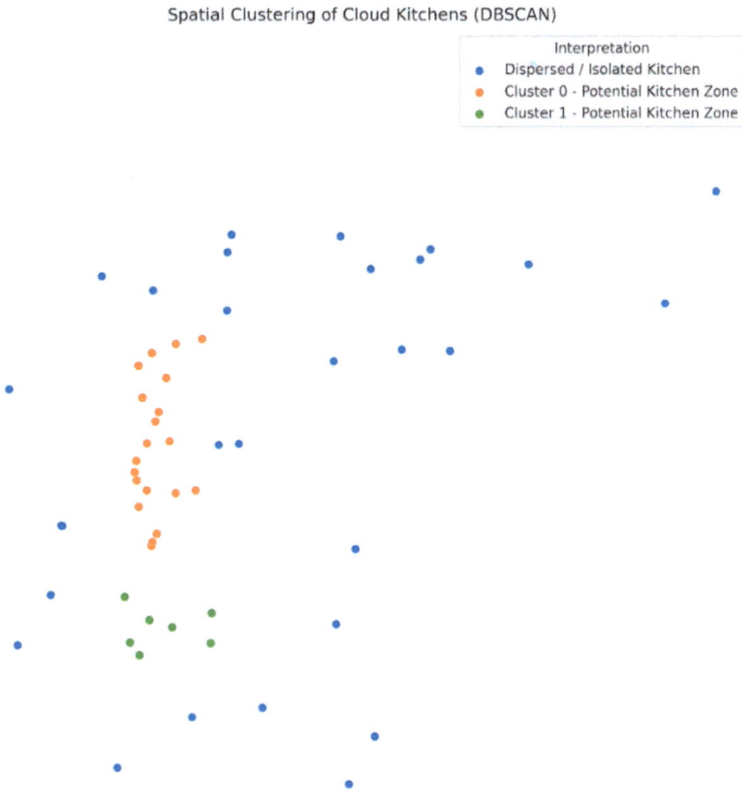

Fig. 14 Spatial clustering of cloud kitchens (DBSCAN). *Source* By Authors

4.2 Platform Economy and Spatial Equity

The overlay of demographic and socioeconomic data with food infrastructure distribution highlights a critical equity dimension. As a reflection of both digital accessibility and purchasing power, cloud kitchens are more common in wards with higher average incomes and literacy rates. On the other hand, despite apparent delivery demand, wards that are peripheral, high-density, and lower-income, like portions of North Kolkata, remain underserved.

This spatial mismatch highlights structural disparities in platform-based food service delivery that reflect larger urban divides in economic opportunity, digital literacy, and infrastructure. The need for focused policy interventions that support inclusive digital urbanism and equitable access to emerging food economies is highlighted by the notable gaps that still exist, even as platforms like Zomato are growing into digitally emerging peripheral zones (such as New Town and Patuli).

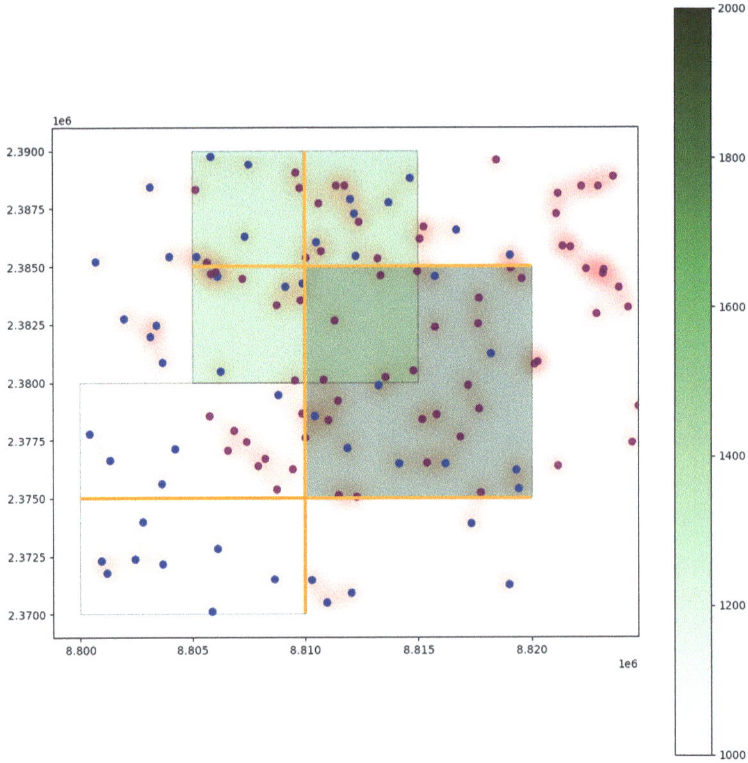

Fig. 15 Catchment area analysis of cloud kitchens in Kolkata overlaid on a KDE heatmap of delivery demand. *Source* By Authors

4.3 Mobility, Delivery Flows, and Urban Rhythms

An unseen logistics layer is actively changing Kolkata's traffic patterns and urban rhythms by identifying high-frequency micromobility corridors that are dominated by two-wheeler deliveries. By highlighting the function of gig-based mobility in supporting new types of localized entrepreneurship and economic agency, the fluidity of these mobility flows challenges conventional ideas of urban transport networks. Such dynamics underscore the necessity for urban planners to incorporate micromobility infrastructures into broader mobility frameworks, addressing issues like congestion management, last-mile connectivity, and road safety within the evolving ecosystem of platform-mediated urban services.

Fig. 16 Cloud kitchen potential score (1–10). *Source* By Authors

4.4 Predictive Modeling and Strategic Planning

A useful tool for predicting future spatial expansions of digital food infrastructure is the ward-wise potential scoring model for cloud kitchen viability. This predictive method allows investors, platform operators, and local government agencies to make data-driven decisions by combining demand, network accessibility, income, and infrastructure factors.

Clusters found along the southern corridors and EM Bypass point to key locations where demand and urban form converge to foster the growth of cloud kitchens. On the other hand, peripheral wards with low scores draw attention to market immaturity or infrastructure bottlenecks, indicating these as longer-term issues that call for coordinated interventions.

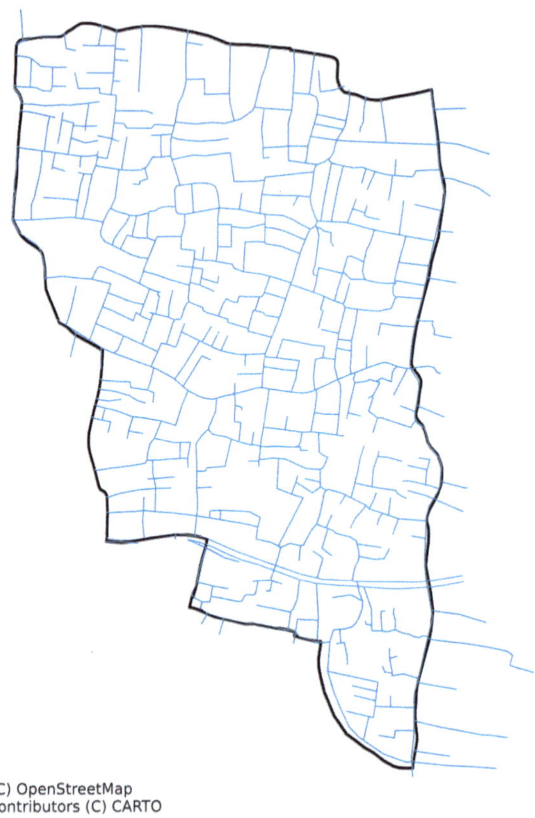

Fig. 17 Road networks ward 105 Score 9.2 (highest potential score). *Source* By Authors

4.5 *Implications for Neo-Futuristic Urbanism*

This study demonstrates how neo-futuristic urbanism, which is defined by elastic, disembodied service layers superimposed on conventional urban morphologies, is produced by platform-mediated food logistics. A paradigm shift in land use, spatial equity, and economic participation is represented by the coexistence of visible and invisible food infrastructures.

Adaptive planning frameworks that strike a balance between innovation and inclusivity are necessary for emerging Indian metropolises to embrace such hybrid digital-physical foodscapes. This is to ensure that the adoption of new technologies does not exacerbate spatial inequality but rather spurs new forms of localized resilience and economic diversification.

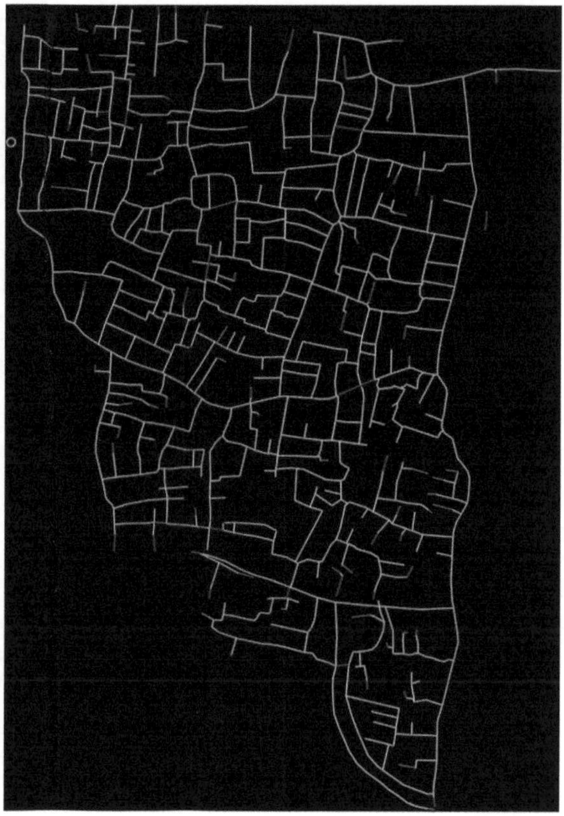

Fig. 18 Connectivity of streets: ward 105. *Source* By Authors

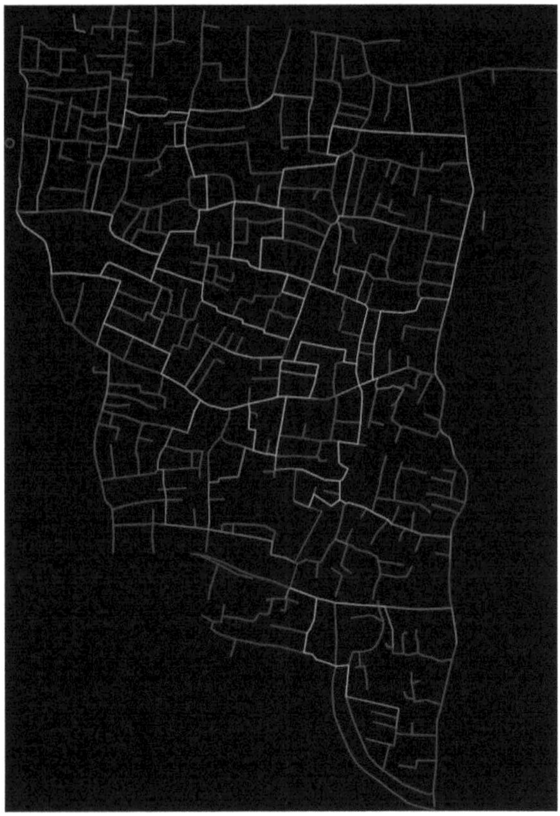

Fig. 19 Choice of street networks: ward 105. *Source* By Authors

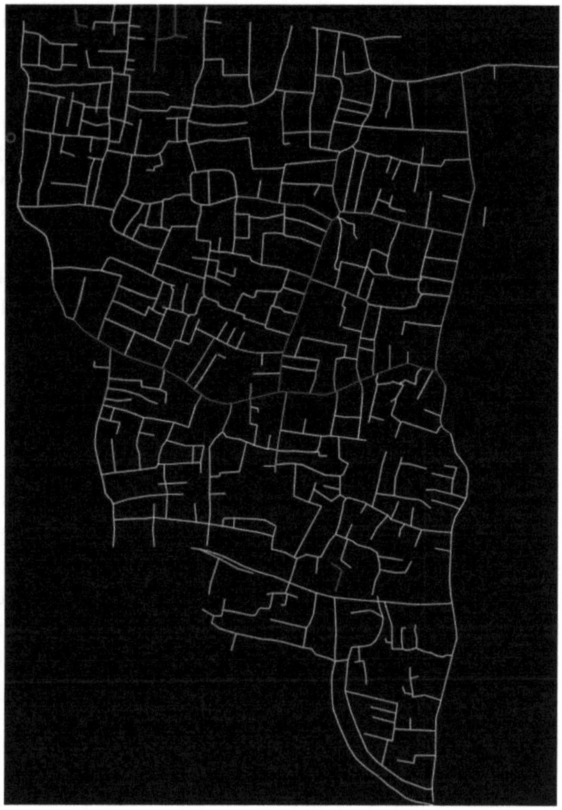

Fig. 20 Integration of street networks: ward 105. *Source* By Authors

5 Conclusion

The primary focus of this study, which examines Kolkata's food infrastructure, is the rise of cloud kitchens and their impacts on demographic, socioeconomic, and spatial factors. It illustrates how the city's culinary scene has undergone significant change as a result of digitalization, consumer behavior, and logistical optimization. The study employs analysis to provide a comprehensive account of how cloud kitchens are transforming the delivery of food in urban areas.

The concentration of cloud kitchens in EM Bypass, Ballygunge, and Kasba suggests a demand-driven urban food network that prioritizes locations with strong digital access, a high residential density, and logistics. These hotspots represent economic opportunity as well as the effectiveness of the infrastructure. On the other hand, the under-representation of cloud kitchens in the high-density, low-income neighborhoods of North Kolkata reveals a digitally divided food economy, suggesting a market gap and a possible area for development. The cloud kitchen viability prediction model uses delivery demand, street integration, income clusters, and accessibility to identify high-opportunity zones like Ward 105. High levels of global and local integration, balanced street selection, and connectedness are confirmed by space syntax analysis, which increases cloud kitchens' ability to service larger catchment areas with fewer redundancies, cut down on delivery times, and lower operational friction.

The analysis of high-scoring wards like Ballygunge and Kasba shows that not all economically successful locations are equally viable. Ballygunge's cellular grid reduces the efficiency of digital delivery models, highlighting the need for a configuration-sensitive model in urban food logistics, particularly in cloud kitchens where demand, income, physical permeability, and city network structure all affect how well digital infrastructure works. This study highlights the importance of food infrastructure in intelligent urban systems. Often invisible, cloud kitchens impact traffic flow, trash management, zoning regulations, and employment patterns. Zoning flexibility, digital access programs, and infrastructure incentives could all be employed as inclusive planning interventions in poor places like Baranagar or Chitpur to encourage the expansion of cloud cooking services.

The evolution of food infrastructure, influenced by digital platforms and spatial conditions, signifies a shift in city function. The gig economy, spatial optimization through platforms like Zomato, and flexible urban entrepreneurship models are redefining the relationship between place, service, and economy, creating a syntactically intelligent future city. Kolkata's cloud kitchen network is a case study of how digital economies spatialize within a city's socio-spatial fabric. It reflects a city in flux, where heritage meets innovation, and the invisible infrastructure of cloud kitchens rewrites urban life. The study advocates for planning paradigms that actively shape this transformation, using tools like space syntax and predictive modeling to create resilient, equitable, and logistically efficient food systems.

6 Way Forward

To better understand the long-term effects of platform urbanism, future research must integrate real-time mobility data, consumer behavior patterns, and policy implications as digital platforms and gig economies continue to shape Kolkata's foodscape. The effects of cloud kitchens on gig economy labor rights, environmental sustainability, and urban planning infrastructure and zoning must be investigated. Urban planners, technologists, and legislators will need to work together to create inclusive, flexible, and equitable urban food systems that leverage digital logistics while reducing socioeconomic and spatial disparities.

7 AI Declaration

This manuscript's Portions were supported through AI-based tools, including advanced language models and spatial analysis algorithms, which assisted in drafting narrative components, synthesizing literature, structuring methodological descriptions, and visualizing analytical outputs. However, all critical judgments, interpretations, and intellectual decisions were independently made by the authors, ensuring the originality and scholarly integrity of the research.

Acknowledgements The authors express their profound gratitude to the Institute for its institutional support and for granting access to critical GIS datasets and analytical resources. Special thanks are extended to the residents and local stakeholders of Kolkata, who graciously participated in the semi-structured interviews, offering invaluable insights into the mobility challenges encountered in everyday life. Their contributions have substantially enhanced the empirical foundation of this research. Furthermore, the authors wish to affirm that all ethical considerations were meticulously followed during the data collection process, ensuring the utmost respect for privacy and establishing informed consent throughout the study. It is noteworthy that no external funding was required for this project.

References

1. Kumar S, Banerji H (2018) Analysis of transformation of urban planning practice by mapping changes in economic, social, cultural and built environment of the neighborhood level urban communities (NLUC): case study of Kolkata, India. In: 54th International society of city and regional planners (ISOCARP) congress 2018, Bodo, Norway. Conference proceedings
2. Mafizul Haque SK, Juber Alam M (2019) Micro-level analysis of road connectivity and its spatial variation in the kolkata municipal corporation (KMC) area. In: Contemporary South Asian Studies, pp 157–176. https://doi.org/10.1007/978-3-030-23796-7_10
3. Antenucci I (2021) Digital infrastructure, liminality, and world-making via Asial infrastructures of extraction in the smart city zones, finance, and platforms in new town Kolkata. Int J Commun 15(0):17. https://ijoc.org/index.php/ijoc/article/view/13973
4. Sinha D, Pandit D (2023) Assessing the economic sustainability of gig work: a case of hyper-local food delivery workers in Kolkata, India. Res Transp Econ 100:101335–101335. https://doi.org/10.1016/j.retrec.2023.101335
5. Caprotti F, Chang I-CC, Joss S (2022) Beyond the smart city: a typology of platform urbanism. Urban Transform 4(1). https://doi.org/10.1186/s42854-022-00033-9
6. Bauriedl S, Strüver A (2020) Platform urbanism: technocapitalist production of private and public spaces. Urban Plan 5(4), 267–276. https://doi.org/10.17645/up.v5i4.3414
7. Raihan MMH, Subroto S, Chowdhury N, Koch K, Ruttan E, Turin TC (2024) Dimensions and barriers for digital (in)equity and digital divide: a systematic integrative review. Digit Transform Soc 4(2). https://doi.org/10.1108/dts-04-2024-0054
8. Hanoon SK, Abdullah AF, Shafri HZM, Wayayok A (2023) Urban growth forecast using machine learning algorithms and gis-based novel techniques: a case study focusing on Nasiriyah city, Southern Iraq. ISPRS Int J Geo Inf 12(2):76. https://doi.org/10.3390/ijgi12020076
9. Rout A, Dawande M, Janakiraman G (2024) Cloud-kitchens: value creation through co-location. Prod Oper Manag. https://doi.org/10.1177/10591478231224950
10. Wei X, Yao XA (2021) Constructing and analyzing spatial-social networks from location-based social media data. Cartogr Geogr Inf Sci 48(3):258–274. https://doi.org/10.1080/15230406.2021.1891974

11. Pan S (n.d.) EPRA Int J Res Dev (IJRD) 8(5). https://doi.org/10.36713/epra2016
12. Muriel JE, Zhang L, Fransoo JC, Perez-Franco R (2022) Assessing the impacts of last mile delivery strategies on delivery vehicles and traffic network performance. Transp Res Part C: Emerg Technol 144:103915. https://doi.org/10.1016/j.trc.2022.103915
13. Shapiro A (2022) Platform urbanism in a pandemic: dark stores, ghost kitchens, and the logistical-urban frontier. J Consum Cult 23(1):146954052110699. https://doi.org/10.1177/14695405211069983
14. Fang C, Chen Z, Liao X, Sun B, Meng L (2024) Urban-rural digitalization evolves from divide to inclusion: empirical evidence from China. NPJ Urban Sustain 4(1). https://doi.org/10.1038/s42949-024-00187-4
15. Joel OT, Oguanobi VU (2024) Data-driven strategies for business expansion: utilizing predictive analytics for enhanced profitability and opportunity identification. Int J Front Eng Technol Res 6(2):071–081. https://doi.org/10.53294/ijfetr.2024.6.2.0035
16. Méndez Domínguez P, Carbonero Muñoz D, Raya Díez E, Castillo J (2023) Digital inclusion for social inclusion. Case study on digital literacy. Front Commun 8. https://doi.org/10.3389/fcomm.2023.1191995
17. Morgan K, Sonnino R (2010) The urban foodscape: world cities and the new food equation. Camb J Reg Econ Soc 3(2):209–224. https://doi.org/10.1093/cjres/rsq007
18. da Cunha DT, Hakim MP, Alves MM, Vicentini MS, Wiśniewska MZ (2024) Dark kitchens: origin, definition, and perspectives of an emerging food sector. Int J Gastron Food Sci 35:100884. https://doi.org/10.1016/j.ijgfs.2024.100884

Smart City as a Catalyst for Social Innovation: The Case of Dubai

Sanjay Nadkarni

Abstract This chapter adopts a conceptual approach to examine how Dubai's smart-city agenda uses urban technologies to address challenges of equity, sustainability, and governance. It draws on theories of social innovation and sociotechnical transitions to frame the analysis. The focus is on the Smart Dubai initiative, which aims to build the "world's smartest and happiest city." Rather than emphasising devices and hardware platforms, the chapter highlights shifting relationships between the public sector, the private sector and society. The analysis examines smart mobility, sustainable urban planning, and data-driven governance. It draws on policy documents, usage data, and case studies to trace how IoT sensors and digital platforms lead to outcomes such as reduced commute times, lower carbon emissions, and higher resident satisfaction. Using a Theory of Change framework and a simplified Social Return on Investment model, the chapter reveals uneven social benefits across different groups. Looking ahead, foresight scenarios identify emerging technologies such as 6G, quantum-secure blockchains, and adaptive regulation as key turning points. The findings suggest that technology delivers public value only when embedded in inclusive and reflexive institutions.

Keywords Dubai · Inclusivity · Smart city · Social impact · Theory of change · Urban technologies

1 Introduction

Global debates on sustainable urban futures increasingly converge on the proposition that digital infrastructures can be mobilised not only to optimise traffic flows or curb energy consumption, but also to confront inequalities that undermine urban

S. Nadkarni (✉)
Plekhanov University, Dubai, United Arab Emirates
e-mail: Nadkarni.S@reu.tech

Thinkers Garden, Vienna, Austria

V. Chand Sharma et al. (eds.), *Tech-Enabled Urbanism and Entrepreneurship for Inclusive Cities*, Lecture Notes in Civil Engineering 779,
https://doi.org/10.1007/978-981-95-3873-7_12

cohesion. Few cities stage this techno-social ambition as prolifically as Dubai. Over little more than four decades the emirate has transformed from a pearl fishing trading port into a hyper-connected metropolis whose skyline and brand identity signal an unrelenting orientation towards technological experimentation. Marketed since 2014 under the banner of becoming the "world's smartest and happiest city," the Smart Dubai programme advances an ambitious claim: that algorithmic governance, ubiquitous sensing and data-driven public services can tangibly improve quality of life for more than 3.6 million residents and twenty million annual visitors [1]. At stake in this experiment is the credibility of the claim echoed in policy circles worldwide that smart-city technologies can serve as platforms for equitable and inclusive urban futures, rather than mere showcases of technocratic efficiencies.

Two interlocking concepts anchor the analysis developed in this chapter. Social innovation is defined as "novel approaches, ideas and organisational structures that address societal challenges and improve quality of life" [2]. The term shifts analytical attention from devices to relationships, emphasising the reconfiguration of power and resources among state, market and civil-society actors. Urban technologies, by contrast, refer to the ensemble of digital networks, information infrastructures and smart devices embedded within the built environment to enhance efficiency, sustainability and residents' well-being [3]. When these two concepts intersect, cities become testbeds for socio-technical experimentation in which hardware and code are entangled with institutional change and new forms of civic participation.

Situated at the maritime crossroads of Asia, Africa and Europe, Dubai has long exploited a strategic geography to cultivate a cosmopolitan, business-friendly image. Building on the oil-fuelled development surges of the 1970s and 1980s, successive UAE rulers pursued diversification through free-trade zones, logistics corridors and globally televised mega-events. The 2014 launch of Smart Dubai represents an inflection in this trajectory, grafting platform-based services, blockchain registries, autonomous-mobility pilots and city-wide connectivity onto earlier place-making logics [4]. By the first quarter of 2025, more than 120 government services had migrated to the DubaiNow super-app, hundreds of kilometres of fibre-optic cable undergirded Internet-of-Things deployments, and pilot zones for flying taxis and robotic delivery vehicles had entered regulatory sandboxes. These developments render Dubai a natural laboratory for examining how urban technologies are mobilised to generate, scale and institutionalise forms of social innovation tailored to a multicultural, rapidly expanding population.

Yet, despite escalating media attention, academic scrutiny of Dubai's smart-city transformation reveals two persistent gaps. First, most studies foreground engineering efficiency or economic competitiveness, leaving under-examined the ways in which digitally mediated services reorganise everyday practices, community relations and urban governance. Second, while issues like digital divides, privacy concerns, and algorithmic bias are acknowledged, discussions often remain speculative and based on Euro-American contexts, with little empirical grounding in regional urban dynamics or Dubai's uniquely diverse population context. Consequently, the mechanisms through which urban technologies might catalyse or constrain social innovation in Dubai remain analytically opaque.

This chapter addresses these gaps through three interrelated objectives.

1. It situates the Smart Dubai vision within theories of social innovation, delineating the socio-technical assemblages through which public value is imagined and operationalised.
2. It interrogates three policy domains (i) digital governance (ii) smart mobility and (iii) sustainable built environment to trace how particular technological deployments translate into organisational change, residents' experience and altered power geometries.
3. It evaluates the implications of these deployments for social inclusion, participatory governance and environmental stewardship, thereby contributing a unique Dubai perspective to comparative smart-city scholarship. In doing so, the chapter tests whether Dubai's model offers transferable lessons or cautionary tales for cities seeking to harness digital transformation as a vehicle for social progress.

The remainder of the chapter is organised as follows. A review of literature in the following section is undertaken to set the context for the ensuing research. Subsequently, the Analysis section elaborates the conceptual framework, drawing on insights from innovation studies, socio-technical transitions and platform urbanism. Impact Assessment provides an assessment of Dubai's flagship initiatives, triangulating policy documents, secondary statistics and illustrative case studies, followed by critical reflections. The future prospects section appraises the social impacts of these initiatives, while emerging technological trajectories in the context of Dubai highlight emergent challenges and opportunities as Dubai navigates the next wave of artificial intelligence, robotics and urban automation. Limitations of this research are acknowledged in the subsequent section followed by the conclusion which distils the main findings and outlines directions for future policy cycles.

By furnishing the theoretical and contextual underpinnings required for a critical interrogation of Dubai's smart-city experiment, this introduction prepares the ground for analysing how urban technologies may impact social innovation in one of the world's most ambitious urban laboratories.

2 Literature Review

Viewed through socio-technical transition theory, urban mobility networks exemplify social innovation by aligning technology, policy, and user behaviour across multiple levels. Geels [5] describes how transforming mobility systems requires co-evolution of physical infrastructures, regulatory regimes, and everyday practices in tandem. Dubai's experience reflects this dynamic. The evolving trends in Dubai's public transit such as the expansion of automated metro and people-mover systems is indicative of niche innovations maturing into a new regime [6]. The city's push for inclusive design within these systems, drawing on strategies like step-free access in metro stations, tactile guidance for the visually impaired, and discounted fares for persons with disabilities is noteworthy [7]. These measures echo global universal design principles and demonstrate social innovation aimed at mobility equity. The importance of behavioural integration via Dubai's multimodal journey-planning app S'hail, which seamlessly links public transport modes has also been gauged [8]. This platform encourages residents to shift routines towards shared mobility, although Yahia et al. note gaps in usage data across different socioeconomic groups that hinder fine-tuned service improvements. Globally, similar integrative efforts are underway: cities like Helsinki have pioneered Mobility-as-a-Service platforms that unify transit options through a single digital interface [9], and Latin American metropolises have innovated with Bus Rapid Transit and aerial cable cars to connect marginalised communities. Such cases illustrate the socio-technical transitions described by theory-physical infrastructure, digital platforms, and user habits coalescing into new, more inclusive mobility regimes [5]. The result is a form of social innovation in urban mobility, wherein technological advances are explicitly steered towards broader social value and accessibility.

In sustainable urban planning, Dubai's initiatives demonstrate the interplay between ecological design and social outcomes. Substantial reductions in carbon emissions and utility bills within Dubai Sustainable City, correlating its solar-powered homes and passive design features with strengthened neighbourhood cohesion have been documented [10]. This planned eco-community, which incentivizes recycling and electric vehicle use, has become a testbed for how environmental sustainability can foster social ties through shared gardens, community farms, and car-free public spaces. Similarly, there is evidence of effectiveness of smart grid upgrades and solar *prosumer* schemes (like the Shams Dubai rooftop solar programme) in the UAE [11, 12]. These sources report improvements in energy efficiency and thousands of buildings equipped with solar panels, yet they also highlight persistent challenges, namely the affordability of green technologies and misaligned incentives between tenants and landlords that slow wider adoption. Such issues are not unique to Dubai. Other cities implementing distributed energy systems have noted similar hurdles in financing and policy, requiring creative solutions [13]. Global case studies reinforce Dubai's experience. For example, the *Vauban* eco-district in Freiburg, Germany achieved dramatic emissions cuts through solar energy, green architecture, and a nearly car-free design, all while cultivating strong social cohesion via resident co-operative planning [14]. Vauban's bottom-up development, groups of citizens (Baugruppen) co-designing their housing clusters, led to both environmental and social benefits, much as Dubai's sustainable community marries low-carbon living with a vibrant communal lifestyle. Dubai Municipality's emerging digital twin initiative operationalizes social and environmental planning criteria in real time, reflecting an emergent paradigm in sustainable urban governance [15]. *Dubai Here*, the city's geospatial digital twin integrates 3D mapping of buildings, land use, and infrastructure with data on energy, mobility, and demographics to inform policy decisions. This enables planners to simulate impacts of new policies or developments on sustainability and social well-being before implementation. Around the world, city governments are exploring similar urban digital twins (e.g. Singapore's Virtual Singapore or Shanghai's MCity) as decision-support systems for scenario testing in climate resilience, land use, and community services [15]. Together, these trends illustrate how sustainable urban planning is increasingly characterised by social innovation leveraging technology (smart grids, solar panels, digital twins) and community engagement to achieve both environmental goals and social co-benefits like inclusivity and cohesion.

In the realm of digital governance, Dubai's innovations are underpinned by co-production theory and data-driven public value frameworks. *Co-production* theory [16] holds that public services are more effective and innovative when citizens are active partners in their design and delivery, rather than passive beneficiaries. Dubai has embraced this approach through initiatives like the Happiness Agenda, which seek to directly involve the public in governance. Measurable efficiency improvements from real-time citizen feedback mechanisms such as the Happiness Meter have been identified [17]. This city-wide tool, deployed across government service centres, invites residents to rate their happiness with services by tapping simple emoticons; policy teams then analyse the live data to tweak services for better satisfaction. By institutionalising a feedback loop, Dubai exemplifies how data-driven co-production can generate public value in practice. Such efforts align with global movements towards "*We-Government*", where digital platforms enable continual collaboration between government and citizens [18]. Beyond feedback tools, Dubai has pursued cutting-edge technologies to enhance transparency and efficiency in government operations. In the context of Dubai's blockchain strategy, the potential of blockchain for public-sector recordkeeping, noting its promise to reduce processing times and fraud through distributed ledger transparency has been explored. In fact, Dubai's leadership launched the "Blockchain Strategy 2020" to shift many transactions (from business registration to visa applications) onto blockchain, aiming to create an ecosystem where paperwork is virtually eliminated, highlighting the government's belief that blockchain can foster both efficiency and trust in public services [19]. Likewise, open data policies have been a cornerstone of Dubai's digital governance framework. An economic impact study of open data, quantifying potential benefits in annual gross value added to the economy projected public value of about 1% of Dubai's GDP illustrates why open data is seen as a driver of innovation, entrepreneurship, and improved services [20]. International experience supports these findings; for example, researchers have noted that open government data can spur new businesses and solutions in fields ranging from transit to healthcare [21]. At the same time, Dubai's integrative data initiatives raise important considerations around privacy and equity. Extensive integration of personal data and city systems, if not coupled with robust cybersecurity and privacy safeguards, can lead to risks such as data breaches or unequal access to digital services [22]. This warning echoes broader critiques in smart city discourse: a "real-time city" awash in sensors and big data analytics may inadvertently reinforce social inequalities or surveillance unless governance frameworks ensure inclusive, ethical use of data [23]. Dubai's policymakers are cognizant of these challenges—initiatives like Dubai's Data Law and the Ethical AI guidelines are steps towards balancing innovation with individual rights. In sum, Dubai's digital governance journey illustrates how co-production and data-driven approaches can create public value (greater efficiency, transparency, economic growth) as forms of social innovation in governance. The city's experience, alongside global examples such as Estonia's national e-government blockchain infrastructure or Barcelona's open data and participatory budgeting platforms, shows both the transformative potential of urban technologies and the necessity of addressing the accompanying governance challenges. By actively involving residents, deploying technology for transparency, and instituting safeguards, Dubai is striving to maximise the social benefits of digital governance while mitigating its risks—a balancing act that many smart cities worldwide are similarly attempting to achieve.

3 Evaluation Framework and Methodology

Anchored in socio-technical transition scholarship, three complementary frameworks were employed to situate and interpret Dubai's urban technology interventions: Theory of Change (ToC) [24], Sustainable Development Goal (SDG) aligned indicators, and Social Return on Investment (SROI) [25].

3.1 Theory of Change (ToC)

ToC is a logic-model approach that explicates the "why and how" behind complex initiatives by articulating the causal chain from resources to impacts. First, inputs (e.g. IoT sensor networks, legislative decrees) are specified alongside underlying assumptions. Next, activities (such as platform integrations or autonomous vehicle pilots) and outputs (for instance, real-time data dashboards or shuttle services) are identified. These lead to short-term outcomes (reduced commute times; elevated satisfaction scores) and, ultimately, long-term impacts (broader social inclusion; lower carbon emissions).

A conceptual representation is provided in Fig. 1. At each link, explicit assumptions are surfaced, such as sufficient digital literacy among end users and potential leakages (where anticipated benefits may not materialise) are noted, enabling transparent tracing of where interventions may falter.

3.2 SDG-Aligned Indicators

To benchmark Dubai's progress against globally recognised targets, three SDG subtargets were adopted:

- SDG 11.2 calls "by 2030 [to] provide access to safe, affordable, accessible and sustainable transport systems for all" with particular focus on vulnerable groups.

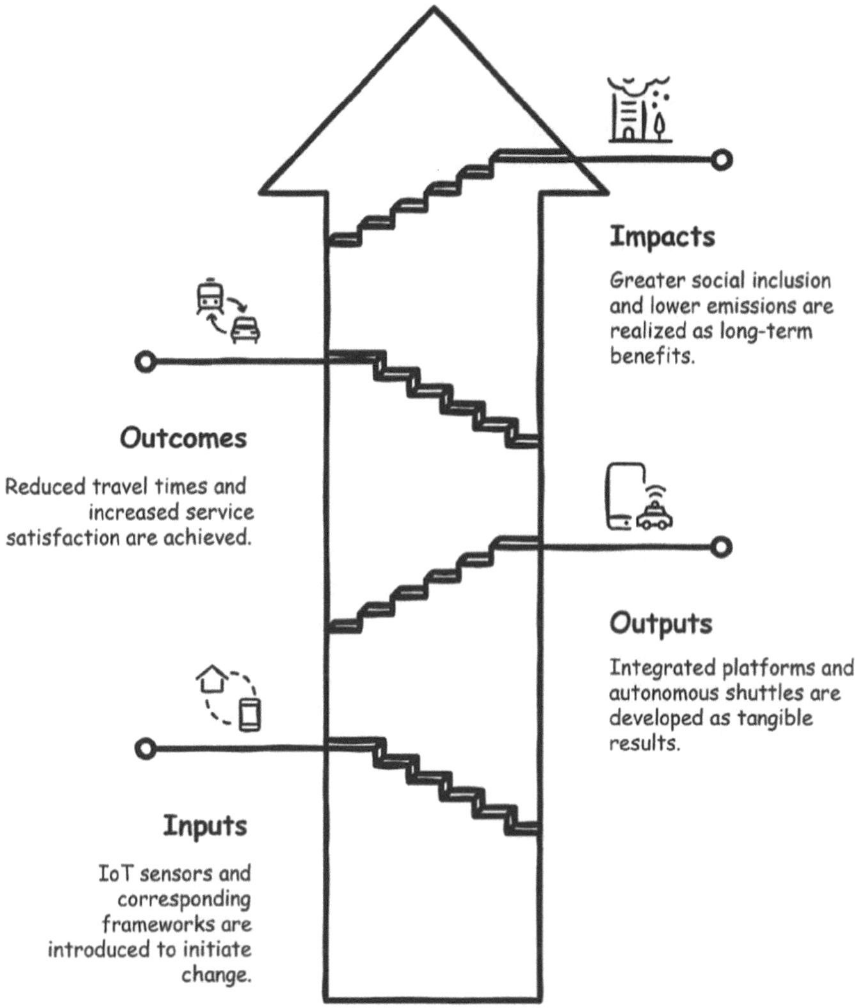

Fig. 1 Theory of change in sustainable urban development. *Source* By author

- SDG 11.6 mandates "reduce the adverse per capita environmental impact of cities," emphasising air quality and waste management.
- SDG 16.6 requires the development of "effective, accountable and transparent institutions at all levels," encompassing measurable benchmarks for public-service responsiveness and transparency.

Each observed metric, e.g. modal share of public transit, per-capita waste generation, citizen satisfaction with e-government services was mapped onto these sub-targets to ensure international comparability and to signal areas of relative strength or underperformance.

3.3 Social Return on Investment (SROI)

SROI is a principle-based method for assigning monetary proxies to social and environmental outcomes that typically lack market prices, thereby enabling their inclusion in a conventional cost–benefit framework. The SROI process as illustrated in Fig. 2 comprises four core steps:

1. Data Collection: Stakeholder-informed gathering of performance and outcome data (e.g. time saved by commuters; reductions in pollutant exposure).
2. Materiality Mapping: Prioritisation of outcomes that are significant to stakeholders and feasible to value.
3. Valuation: Application of financial proxies such as average wage rates for travel-time savings or healthcare cost averages for improved air quality to convert outcomes into monetary terms.
4. Calculation: Aggregation of monetized benefits and comparison against documented intervention costs to yield a benefit-to-cost ratio.

Crucially, SROI mandates stakeholder engagement across all stages to ensure that the values most meaningful to affected communities are captured, thereby grounding the analysis in local perspectives and counteracting bias. However, as this research is conceptual by design, direct stakeholder consultations were not conducted; instead, the analysis relied on secondary data and literature to represent stakeholder perspectives. This is noted as a limitation of the SROI assessment.

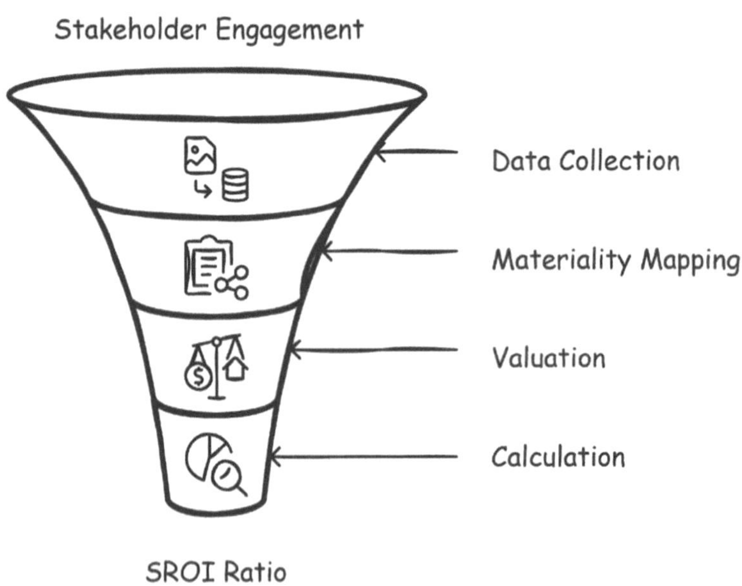

Fig. 2 SROI process funnel. *Source* By author

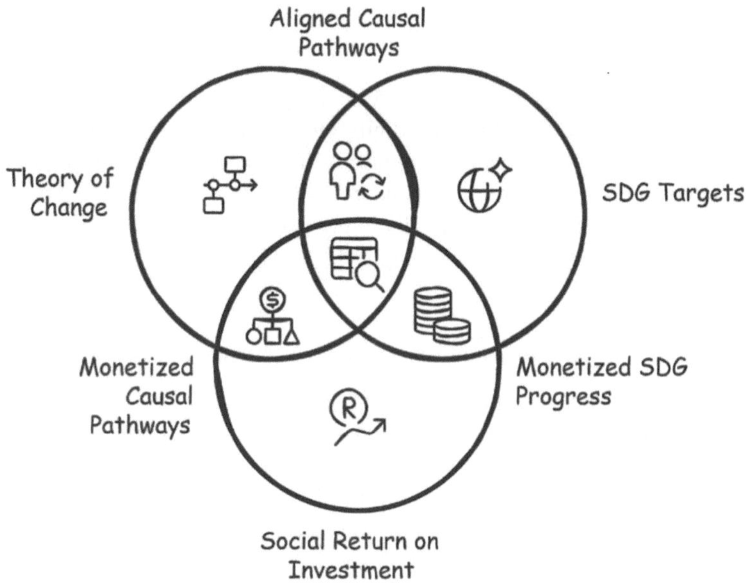

Fig. 3 Integrated impact assessment of urban technology. *Source* By author

As illustrated in the stakeholder engagement funnel represented in Fig. 2, materiality mapping ensures that the context is factored, thereby minimising bias. Consequently, the valuation which serves as an input to the calculation of the SROI ratio tends to be more reliable as it is grounded in the local perspectives.

By weaving together ToC's detailed causal mapping, SDG targets' standardised benchmarks, and SROI's monetized valuation of social value, the methodology delivers a robust, multi-angled appraisal of how Dubai's urban technology initiatives generate, and where they may fall short of, intended social, economic, and environmental outcomes. This integrated framework is represented diagrammatically in Fig. 3.

Data sources span civic dashboards (RTA Open Data, Digital Dubai's Happiness Meter API, DEWA's smart-metre portal), secondary surveys [26] survey)), and policy evaluation reports [1, 27, 28]. Where official datasets remain proprietary, estimates are triangulated from peer-reviewed studies [4].

Constraints include short time-series baselines (post-2014), incomplete demographic disaggregation, and vendor-reported metrics awaiting independent audit. These limitations temper causal claims but do not invalidate directional findings.

3.4 Research Design

This chapter adopts a single-case, embedded design with Dubai as the overarching case and mobility, sustainability, and digital governance as embedded units, suitable for theory-informed conceptual inquiry [29]. The research follows a conceptual, mixed-method logic in which secondary quantitative indicators and qualitative documentary evidence are systematically integrated to answer the guiding questions on social innovation and urban technologies [30, 31].

Data collection

Data collection was through secondary sources. A targeted document analysis of:

(a) official policy and evaluation reports [1, 27, 28, 32, 33]
(b) administrative/open-data portals (Digital Dubai Happiness Meter API; DEWA smart-metre portal; DSC household travel surveys 2019–2023)
(c) peer-reviewed literature on smart cities, SROI, and socio-technical transitions. Sources were identified through keyword searches ("Dubai smart city", "social innovation", "SROI", "digital governance", "mobility equity") across Scopus, Web of Science, Research Gate and Google Scholar with inclusion criteria of relevance to Dubai (or comparable geographical/metropolitan contexts), methodological transparency, and publication between 2014–2025.

Analytical Techniques

Three complementary lenses were adopted for the purpose of analysis of the data collated through the hitherto described sources:

1. **Theory of Change (ToC)**: Outcomes chains were mapped by linking inputs/activities to outputs, short-term outcomes, and long-term impacts, while making provisions for assumptions and potential leakages [24].
2. **SDG Indicator Alignment**: Observed metrics (e.g. modal share, per-capita waste, service satisfaction) were mapped to SDG sub-targets 11.2, 11.6, and 16.6 to benchmark progress and gaps.
3. **Social Return on Investment (SROI)**: Following established guidance [25, 34], outcomes material to stakeholders were indicatively gauged for directionality using proxies (e.g. average wage rates for time savings, healthcare cost estimates for pollution reduction) in the absence of the required data to compute actual values.

Validity and reliability. Credibility was enhanced through triangulation (cross-checking figures across multiple documents), explicit documentation of assumptions, and alignment with recognised frameworks (ToC, SDGs, SROI). Where data gaps existed (e.g. demographic disaggregation), these were flagged and claims qualified accordingly, rather than over-extending inference. This audit trail aims to mitigate bias and strengthen inferential rigour [35].

3.5 *Machine Dependency of Methodology*

In today's evolving research landscape, the integration of generative AI tools into academic workflows has become not only prevalent but, increasingly, a matter requiring explicit ethical transparency. This conceptual study is no exception. In alignment with emerging norms for human–machine collaboration in scholarship, this paper adopts the framework for research and publication standards introduced by the Dubai Future Foundation [33], as illustrated in Fig. 4.

Figure 4 serves as a visual guide, offering an ordinal scale to map the degree of generative AI's (machine) contribution across different phases of the research process.

Application of this framework delineates the extent of human versus AI involvement specific to this study. Core elements of the research which include ideation, literature synthesis, data gathering, and narrative construction were led by the author (human), with generative AI tools (machine) playing a supportive role confined to factual verification and refinement of the written output. The interpretation of sourced data, drawn from an array of referenced literature was undertaken solely by the author. Given the absence of primary data analysis, the role of AI in this dimension was minimal to non-existent. Conversely, Figs. 1, 2, and 3 were conceptualised by the author and the rendering was executed by AI tools. The author's input in these instances was to craft textual prompts that defined the conceptual and contextual parameters of each illustration.

By clarifying this methodological approach including the nature and scope of AI tool usage, this paper adheres to principles of research integrity and transparency

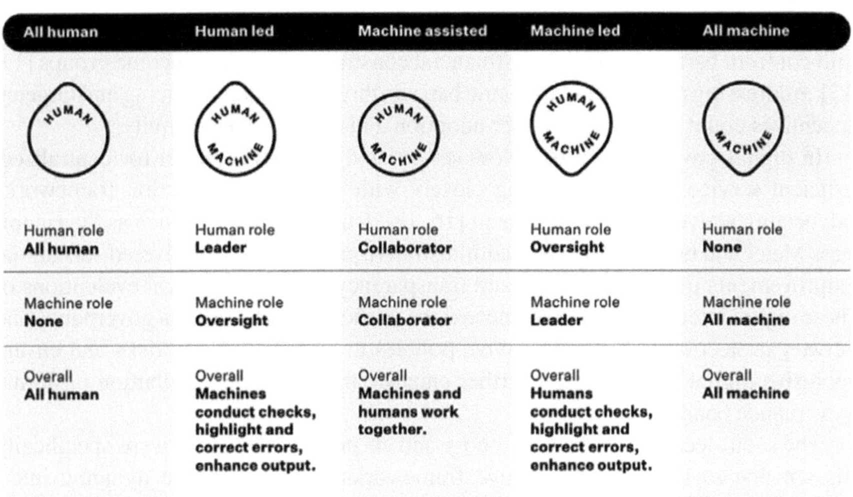

Fig. 4 Human–machine research collaboration scale. *Source* Dubai Future Foundation

in digital scholarship. The following section proceeds to analyse the secondary data corpus, as structured through this collaborative human–machine framework.

4 Analysis

Dubai's approach to integrating smart transportation, sustainable urban planning, and data-driven digital governance demonstrates how socio-technical strategies translate ambitious policy goals into tangible social outcomes. Drawing from socio-technical transition theory [5], this analysis assesses how effectively Dubai deploys integrated technologies to foster inclusive social innovation.

In transportation, Dubai's automated metro and multimodal platform, S'hail, reflect strategic initiatives that significantly reduce congestion and emissions while promoting inclusive mobility [1, 27]. The introduction of autonomous transit systems aligns well with broader global trends, contributing not only to environmental outcomes but also to enhanced labour market opportunities. However, the full potential for equitable service delivery remains partially unrealized due to limited detailed demographic usage data [8]. Ensuring comprehensive data harvesting and dissemination could further optimise inclusivity and accessibility of Dubai's mobility services.

Sustainable urban planning efforts in Dubai, epitomised by Dubai Sustainable City, demonstrate that integrated ecological design effectively promotes social cohesion while achieving significant reductions in residents' carbon footprints [10]. Similarities with international benchmarks like Freiburg's Vauban district illustrate the global relevance of community-driven sustainability initiatives. Nonetheless, Dubai's sustainability programmes, including smart grids and renewable energy initiatives, still confront barriers, particularly financial constraints for lower-income groups [11, 12]. Addressing these socio-economic barriers through equitable pricing and targeted incentives could accelerate broader adoption and enhance social equity.

In digital governance, DubaiNow serves as a successful model for centralised, efficient service delivery, aligning closely with global co-production frameworks advocating active citizen engagement [16, 18]. Dubai's initiatives such as the Happiness Meter and blockchain-based administrative processes have delivered substantial improvements in responsiveness and transparency. However, critical evaluations of these digital interventions remain necessary, particularly around data governance and privacy protection [22, 23]. Proactive policies to mitigate privacy risks and ensure inclusive digital literacy could further enhance the equitable distribution of digital governance benefits.

The socio-technical transition theory and co-production theory were specifically chosen due to their comprehensive frameworks that capture the dynamic interplay between technology, institutional frameworks, and societal practices, all essential elements in understanding Dubai's innovative urban context. Socio-technical transition theory effectively explains the interconnectedness and co-evolution of infrastructure, regulations, and user practices, providing a holistic view suitable for analysing large-scale urban transitions [5]. Co-production theory complements this by emphasising active citizen engagement in governance processes, crucial for ensuring responsiveness and legitimacy in technological implementations [16, 18].

Alternative theoretical frameworks that could further enrich analysis of social innovation through urban technologies in Dubai include the Actor-Network Theory (ANT) and Social Practice Theory. ANT could provide valuable insights into how technology and human interactions co-create urban environments, emphasising the role of technology as an active participant rather than a passive tool [36]. Social Practice Theory, on the other hand, emphasises routines and daily practices, illuminating how residents adopt and adapt urban technologies in everyday life, potentially revealing new strategies to enhance social acceptance and equitable usage [37].

Having traced how Dubai's mobility, urban-form and governance initiatives operationalise the emirate's smart-city vision, the next step is to determine whether these techno-social assemblages are in fact delivering the equitable, sustainable and human-centred benefits their architects proclaim. The ensuing Impact Assessment therefore subjects the empirical cases to a structured appraisal of outcomes, isolating both demonstrable gains and residual gaps that call for policy recalibration.

5 Impact Assessment

This section applies the integrated framework outlined in the preceding sections incorporating **Theory of Change (ToC), SDG-aligned indicators,** and a **simplified Social Return on Investment (SROI)** to analyse Dubai's smart mobility, sustainable urban planning, and digital governance initiatives. The aim is to make causal chains explicit, benchmark outcomes against globally recognised targets, and indicate monetisable public value, while highlighting assumptions and leakages that impact inclusivity claims [24, 25].

5.1 Smart Mobility (SDG 11.2: Safe, Affordable, Accessible, and Sustainable Transport for All)

ToC application: Inputs/activities- automated metro expansion, the S'hail multimodal journey-planning app, and AI-enabled intelligent transport systems-produced outputs such as real-time routing and optimised signal control. These, in turn, yielded outcomes of reduced average commute times and increased public transport ridership. Critical ToC assumptions include sufficient digital literacy and spatial coverage; leakages are evident where peripheral labour camps depend on informal minibuses and where autonomous pilots cluster in affluent zones [5, 8, 27, 28].

Observed outcomes and SDG mapping. Mean commute time declined from 41 min in 2014 to 34 min in 2023, with ITS optimisation contributing an additional 6%

reduction [28]. These gains directly support SDG 11.2's accessibility and efficiency targets [38].

SROI signal. Time savings translate into labour-productivity gains, one of the largest monetisable benefits in urban mobility SROI studies, though full cost data for comprehensive ratio calculation are absent. A conservative interpretation therefore reports directional indicator rather than a headline SROI figure [25].

5.2 Sustainable Urban Planning (SDG 11.6: Reduced Per-capita Environmental Impact)

ToC application: IoT-enabled smart grids, shaded pedestrian infrastructure, and eco-district design (e.g. Dubai Sustainable City) function as inputs/activities that generated outputs such as peak-load curtailment, passive cooling, and car-free public areas. Outcomes include lower per-capita energy intensity and higher street-level activity. Assumptions hinge on affordability of green technologies and alignment of tenant–landlord incentives; leakages are evident where lower-income groups face adoption barriers [10, 11, 15].

Observed outcomes and SDG mapping. Energy intensity per capita fell 12% since, Sustainable City achieved an ~ 85% emissions reduction, and shaded streets reported ~ 35% higher footfall than conventional suburbs [10, 12]. These indicators align with SDG 11.6 on environmental impact mitigation.

SROI signal. Avoided carbon and health costs, plus enhanced liveability, are monetisable streams. However, reliance on vendor-reported baselines and proprietary datasets constrains precise valuation. Directionality is therefore emphasised pending independent audits and fuller disclosure.

5.3 Digital Governance (SDG 16.6: Effective, Accountable, and Transparent Institutions)

ToC application. Consolidation of services in the DubaiNow super-app, real-time customer feedback via the Happiness Meter, and blockchain-based registries served as inputs/activities producing outputs such as streamlined transactions and transparency dashboards. Outcomes include improved satisfaction scores and drastic reductions in processing times. A key assumption-universal digital uptake is only partially met: DubaiNow penetration stands at ~ 78% overall but drops to 54% for residents over 55 and 47% for those with primary education, signalling equity leakages [17, 39].

Observed outcomes and SDG mapping. Happiness scores rose from 86.5% (2015) to 90.2% (2024), though districts like Deira remain below 80% [40]. Blockchain reduced property-transfer processing to under one minute, saving an estimated AED 5.5 million in labour costs [19]. These results advance SDG 16.6's transparency and efficiency targets.

SROI signal. Efficiency savings and trust gains are monetisable; nevertheless, in the absence of audited life-cycle costs for the digital stack, a definitive SROI ratio computation is not possible. An independent verification to strengthen evidentiary claims would be necessary [20].

5.4 *Cross-Domain Synthesis*

Across domains, benefits concentrate along data-rich corridors, while digitally or spatially marginalised groups accrue fewer gains eroding aggregate SROI and challenging ToC assumptions about inclusive pathways [22, 23]. Economically, smart-city projects added an estimated AED 14 billion to GDP in 2023 (60% from ICT subsectors), yet conflicting job-creation counts-1,800 (RTA) versus 1,200 (Dubai Statistics Centre), reflect divergent definitions (direct vs. indirect, temporary vs. permanent). Therefore employment impacts need to be interpreted with a high degree of caution, pending harmonised metrics and transparent methodologies in the future audits.

Overall, the integrated framework reveals *directionally positive* social value, but its realisation is conditional on closing demographic data gaps, widening digital literacy, and embedding independent audits into valuation practices. Technology delivers public value when coupled with inclusive institutions and reflexive evaluation, precisely where Dubai's next policy cycle must concentrate.

6 Critical Reflections

Benefits cluster along data-rich corridors, leaving digital-marginalised groups behind. Privacy and algorithmic transparency remain under-governed. Policy remedies include targeted digital-literacy hubs, data-trust audits, and revenue-linked equity fundsViewed through a Theory-of-Change lens, the inputs of Dubai's smart-city programme, i.e. accessibility upgrades to DubaiNow, ITS algorithms, shaded pedestrian infrastructure, generate immediate outputs such as a high platform penetration rate and integrated, real-time mobility services; these in turn produce measurable outcomes: a perceptible fall in mean commute times since 2014, an approximate four-point rise in reported happiness, and demonstrably higher street-level activity in shaded districts.

The simplified Social Return on Investment frames these outcomes in monetised public value: time savings alone translate into labour-productivity gains, while enhanced well-being and reduced carbon exposure lower future health-care costs, together contributing to the GDP boost attributed to smart-city projects in 2023. Yet ToC also highlights the broken causal links for marginalised groups: only half of residents over 55 or with primary education leverage DubaiNow, and labour camps remain outside formal transit networks, signalling leakage that erodes aggregate SROI. Thus, the social dividend is significant but contingent on closing digital-literacy and service-coverage gaps so that each step in the causal chain delivers equitable value across the population.

Across domains, technology delivers social value only when paired with inclusive governance and continuous evaluation. The Future Prospects section probes how emerging trajectories such as 6G, quantum-secure blockchains, adaptive regulation might recalibrate these dynamics in the foreseeable future.

7 Future Prospects

While the preceding Impact Assessment traced the tangible social, economic and environmental outcomes of Dubai's smart-city interventions to date, a comprehensive understanding of their long-term significance requires a forward-looking lens. Consequently, this section probes plausible future pathways (technological, institutional and societal) that are likely to shape the emirate's quest for inclusive, sustainable innovation over the next decade. By combining structured foresight techniques with evidence from official strategy documents, it elucidates the contingencies, risks and opportunities that will condition Dubai's smart-city trajectory in the coming decade.

A three-stage horizon-scanning protocol that employs a STEEP matrix, and trend-impact analysis could be a way forward to construct three scenarios: Incremental Optimisation, Platform Convergence and Resilient Regeneration [32]. The horizon is 2035, aligning with the Dubai Economic Agenda D33 and UAE Net Zero 2050. Key drivers include demographic ageing, heat-stress escalation, edge-AI maturation and data-sovereignty norms. Limitations involve path-dependence, sparse long-run data, and qualitatively treated wild cards.

7.1 Emerging Technological Trajectories

Emerging technologies and policy innovations are rapidly reshaping the landscape of digital governance, service delivery, and sustainability. As the UAE advances its strategic ambitions under frameworks like D33, the integration of emerging technologies introduces both unprecedented opportunities and complex risks. Concurrently, adaptive regulatory mechanisms, cross-emirate policy harmonisation, and climate-aligned economic instruments are being explored to ensure inclusivity, resilience, and equity. The following points outline key developments and their implications for public sector transformation and institutional futures.

- 6G and ubiquitous sensing: Terahertz bandwidth and millisecond latency could enable hyper-personalised public services while intensifying cybersecurity risks [41].
- Quantum-secure blockchains: Hybrid chains layering post-quantum cryptography may future-proof land-registry ledgers but raise transaction costs [19].
- Edge AI and federated learning: RTA pilots cut bandwidth by 70% but shift governance to opaque model weights.
- Institutional and Policy Futures: Adaptive regulation and data trusts. Multi-sector sandboxes with embedded ethical triggers and fiduciary data trusts could balance experimentation and privacy [32].
- Cross-emirate harmonisation: D33 success hinges on interoperable labour and identity policies [42].
- Climate alignment: Carbon budgets in KPI dashboards and congestion-pricing-funded just-transition schemes can link decarbonisation and equity [43].

7.2 Societal and Environmental Implications

Ageing populations will expect predictive services and accessibility by default. Automation risks polarising incomes unless there is significant investment in re-skilling [44]. Heat projections of + 2 °C by 2050 demand digital-twin-guided adaptation, yet financing gaps persist [45]. Dubai's smart-city trajectory hinges on quantum-secure infrastructures, adaptive governance and demographic shifts. Realising an inclusive path requires iterative regulation, equitable financing and robust evaluation.

8 Limitations and Scope for Future Research

Findings are bounded by short post-2014 data, incomplete demographic disaggregation and scenario uncertainty. Dubai's fiscal capacity and governance structure may limit transferability to cities with fewer resources or different institutional contexts. Given the conceptual nature of this research, actual computation of the impact assessment metrics was not feasible on account of the lack of primary data. Factoring these limitations, following is the suggested agenda for future research.

1. Longitudinal well-being studies linking sensor data to household surveys can clarify capability gains over time.

2. Comparative cross-emirate analyses will test scalability of Dubai's innovations within diverse UAE contexts.
3. AI ethics inquiries should examine how federated learning and quantum-secure infrastructures reshape accountability for bias and energy consumption.

Having acknowledged the limitations of this study and the scope for future research, the Conclusion distils the insights developed in this paper into actionable recommendations.

9 Conclusion and Closing Synthesis

Having mapped plausible technological, institutional and societal trajectories to 2035, the chapter now turns to a summative appraisal. The ensuing conclusion distils principal insights into actionable guidance, acknowledges methodological shortfalls, and charts research pathways capable of sharpening Dubai's social-innovation agenda and informing global smart-city debates.

This chapter investigated how urban technologies intersect with social innovation in Dubai's quest to become "the world's smartest and happiest city." It traced causal chains from metro automation and digital-twin planning to measurable gains in mobility, sustainability and governance, while highlighting persistent equity gaps. Foresight scenarios showed that 6G, quantum-secure blockchains and adaptive regulation could either amplify or hinder inclusivity and resilience [32]. The analysis demonstrates that technology yields social value only when embedded in equitable institutions, continuous evaluation and human-centred design. Therefore, the actionable recommendations for policy makers and practitioners are as follows:

1. Institutionalise adaptive governance sandboxes for emerging technologies to test quantum-secure data trusts with embedded bias, privacy and carbon metrics [28].
2. Deploy fiduciary data-trust frameworks that grant residents enforceable rights over municipal data reuse, mitigating surveillance concerns [22].
3. Expand multilingual digital-literacy programmes and subsidised connectivity to bridge the 20-point adoption gap among older and low-income residents [39].
4. Link congestion-pricing revenues to a just-transition fund allocating 30% towards autonomous-shuttle subsidies and rooftop-solar rebates in under-served districts [43].
5. Mandate third-party audits of Happiness Meter scores, blockchain ledgers and AI traffic algorithms to strengthen evidence quality and public trust [32].

Dubai's smart-city journey illustrates how tightly coupling digital infrastructure with an ambitious urban vision can generate tangible improvements in everyday life. Integrated mobility, frictionless e-government and data-driven energy management have begun to recalibrate the relationship between residents, technology and the city, reinforcing the narrative that purposeful innovation can serve the public good.

Yet the same transformation surfaces unresolved questions of equity, governance and accountability. Digital adoption remains uneven across socio-economic and resident strata; blue-collar labour settlements and informal neighbourhoods still operate at the periphery of smart-city logics; and proprietary algorithms continue to mediate key urban decisions with limited transparency. These factors could trigger latent leaks in the Theory-of-Change where safeguards, affordability mechanisms and participatory design have not kept pace with technical rollout.

The next policy cycle must therefore rest on three conceptual pillars.

(i) Adaptive governance calls for regulatory frameworks that evolve in tandem with emergent technologies while embedding privacy, fairness and environmental stewardship by design.
(ii) Inclusive capability building demands sustained investment in digital literacy, accessible user interfaces and redistributive mobility models to ensure no community is left behind.
(iii) Reflexive evaluation requires systematically pairing rigourous causal-inference techniques with qualitative insight so that societal impacts are continuously verified and course-corrections become a feature, not a remedy.

This chapter's exploratory analysis should be read as a waypoint, inviting deeper longitudinal inquiry that blends sensor-level telemetry with lived-experience narratives and comparative benchmarking with peer cities in the region and beyond. Sustaining the promise of the smart-city ideal ultimately hinges on the human-centred approach to innovation and technology interventions which prioritises people over platforms, so as to serve as a catalyst for collective flourishing, and not an end in itself.

In conclusion, Dubai's experience illustrates that smart-city technologies can unlock significant social value, but only through adaptive regulation, inclusive design and transparent evaluation. By operationalising these recommendations and pursuing targeted research, the emirate and other aspiring smart cities can steer digital transformation towards genuinely equitable and sustainable urban futures.

References

1. Dubai Future Foundation (2025) Future trends: transportation. https://www.dubaifuture.ae/wp-content/uploads/2020/08/DFF-Covid19-Transportation-ENG.pdf. Last accessed 28 Jul 2025
2. Mulgan G, Tucker S, Ali R, Sanders B (2025) Social innovation: what it is, why it matters, how it can be accelerated. https://youngfoundation.org/wp-content/uploads/2012/10/Social-Innovation-what-it-is-why-it-matters-how-it-can-be-accelerated-March-2007.pdf. Last accessed 28 Jul 2025
3. Batty M (2013) The new science of cities. MIT Press
4. Virtudes A, Abbara A, Sá J (2017) Dubai: a pioneer smart city in the Arabian territory. IOP Conf Ser-Mater Sci Eng 245(5):052071
5. Geels FW (2012) A socio-technical analysis of low-carbon transitions: introducing the multi-level perspective into transport studies. J Transp Geogr 24:471–482

6. Srivastava SD, Agrawal R (2014) Automated people movers: a futuristic approach to modern transportation planning. IOSR J Mech Civ Eng 11(3):1–11
7. Pineda VS (2020) Building the Inclusive city: governance, access, and the urban transformation of Dubai. Springer Nature
8. Yahia O, Chohan AH, Arar M, Awad J (2025) Toward sustainable urban mobility: a systematic review of transitoriented development for the appraisal of Dubai metro stations. Smart Cities 8(1):21
9. Sochor J, Arby H, Karlsson IM, Sarasini S (2018) A topological approach to mobility as a service: a proposed tool for understanding requirements and effects, and for aiding the integration of societal goals. Res Transp Bus Manag 27:3–14
10. Juaidi A, AlFaris F, Saeed F, SalmerónManzano E, ManzanoAgugliaro F (2019) Urban design to achieve sustainable energy of residential neighbourhoods in arid climates. J Clean Prod 228:135–152
11. Sahib U (2019) Smart Dubai: sensing Dubai smart city for smart environment management. In: Vinod Kumar TM (ed) Smart environment for smart cities. Springer, Singapore, pp 437–489
12. DEWA Smart meter programme annual report. https://www.dewa.gov.ae/en/about-us/media-publications/latest-news/2024/02/dubai-electricity-and-water-authority-pjsc-announces. Last accessed 28 Jul 2025
13. OECD (2019) Accelerating climate action: refocusing policies through a well-being lens. https://doi.org/10.1787/2f4c8c9a-en. https://www.oecd-ilibrary.org/environment/accelerating-climate-action_2f4c8c9a-en. Last accessed 28 Jul 2025
14. Coates GJ (2013) The sustainable urban district of Vauban in Freiburg, Germany. Int J Des Nat Ecodynamics 8(4):265–286
15. Shahat E, Hyun CT, Yeom C (2021) City digital twin potentials: areview and research agenda. Sustainability 13(6):3386
16. Ostrom E (1996) Crossing the great divide: coproduction, synergy, and development. World Dev 24(6):1073–1087
17. Zakzak L (2019) Citizencentric smart city development: the case of smart Dubai's "happiness agenda". In: Proceedings of the 20th annual international conference on digital government research (dg.o 2019). ACM, pp 141–147
18. Linders D (2012) From e-government to we-government: defining a typology for citizen co-production in the age of social media. Gov Inf Q 29(4):446–454
19. Bishr AB (2019) Dubai: a city powered by blockchain. Innov: Technol, GovAnce, Glob 12(3–4):4–8
20. Smart Dubai Office & KPMG. Dubai Open and Shared Data Economic and Social Impact Study—Final Report. https://www.digitaldubai.ae/docs/default-source/publications/dubai-open-and-shared-data-economic-and-social-impact-study.pdf. Last accessed 28 Jul 2025
21. Janssen MFWHA, Charalabidis Y, van Eijk AMG (2012) Benefits, adoption barriers and myths of open data and open government. Inf Syst Manag 29(4):258–268
22. Tubaishat A, Al Jouhi M (2020) Building a security framework for smart cities: a case study from UAE. In: Proceedings of the 5th international conference on computer and communication systems (ICCCS 2020). IEEE, pp 477–481
23. Kitchin R (2014) The real-time city? Big data and smart urbanism. GeoJournal 79(1):1–14
24. Connell JP, Kubisch AC (1998) Applying a theory of change approach to the evaluation of comprehensive community initiatives: Progress, prospects, and problems. In: Fulbright-Anderson, K, Kubisch AC, Connell JP (eds) New approaches to evaluating community initiatives. Aspen Institute 2:15–44
25. Lingane A, Olsen S (2004) Guidelines for social return on investment. Calif Manag Rev 46(3):116–135
26. Dubai Statistics Centre (2024) Statistical yearbook. https://www.dsc.gov.ae/. Last accessed 28 Jul 2025
27. Roads and Transport Authority (2022) RTA endorses 5year sustainability plan 2021–2025. https://mediaoffice.ae/en/news/2022/April/17-04/RTA-endorses-5-Year-Sustainability-Plan-2021. Last accessed 28 Jul 2025

28. Roads and Transport Authority (2025) Traffic signal control system upgrade using AI and digital twin technology. https://www.rta.ae/wps/portal/rta/ae/home/news-and-media/all-news/NewsDetails/launching-traffic-signal-control-system-upgrade-using-ai-and-digital-twin-technology. Last accessed 28 Jul 2025
29. Yin RK (2018) Case study research and applications: design and methods, 6th ed. Sage
30. Jaakkola E (2020) Designing conceptual articles: four approaches. AMS Review 10(1–2):18–26
31. Snyder H (2019) Literature review as a research methodology: an overview and guidelines. J Bus Res 104:333–339
32. Dubai Future Foundation (2025) The Global 50: key trends shaping the future of Dubai and the world https://www.dubaifuture.ae/wp-content/uploads/2023/04/THE-GLOBAL-50-EN.pdf. Last accessed 28 Jul 2025
33. Dubai Future Foundation (2025) Framework for research and publications standards for human-machine collaboration. https://www.dubaifuture.ae/wp-content/uploads/2025/07/HMC-White-paper-ENG.pdf. Last accessed 28 Jul 2025
34. Social Value International (2021) SROI guide and standards https://socialvalueint.org. Last accessed 28 Jul 2025
35. Tranfield D, Denyer D, Smart P (2003) Towards a methodology for developing evidence-informed management knowledge by means of systematic review. Br J Manag 14(3):207–222
36. Latour B (2005 Reassembling the social: an introduction to actor-network-theory. Oxford University Press
37. Shove E, Pantzar M, Watson M (2012) The dynamics of social practice: everyday life and how it changes. Sage Publications
38. UNDP (2025) Transforming our world: the 2030 agenda for sustainable development https://www.undp.org/ukraine/publications/transforming-our-world-2030-agenda-sustainable-development. Last accessed 28 Jul 2025
39. Digital Dubai (2025) Our initiatives https://www.digitaldubai.ae/initiatives. Last accessed 28 Jul 2025
40. Dubai Protocol Department. Hamdan bin Mohammed approves results of 2024 Dubai Government Customer and Employee Happiness Indices, Mystery Shopper Survey. https://www.protocol.dubai.ae/en/media-listing/news-events/hamdan-bin-mohammed-approves-results-of-2024-dubai-government-customer-and-employee-happiness-indices-mystery-shopper-survey/. Last accessed 28 Jul 2025
41. International Telecommunication Union (2025) Framework and overall objectives of the future development of IMT for 2030 and beyond (Recommendation ITU-R M.2160–0) https://www.itu.int/dms_pubrec/itu-r/rec/m/R-REC-M.2160-0-202311-I!!PDF-E.pdf. Last accessed 28 Jul 2025
42. Government of Dubai. Dubai Economic Agenda D33. https://u.ae/en/about-the-uae/strategies-initiatives-and-awards/strategies-plans-and-visions/finance-and-economy/dubai-economic-agenda-d33. Last accessed 28 Jul 2025
43. Ministry of Climate Change and Environment (2025) UAE Net zero 2050 strategic initiative. https://u.ae/en/about-the-uae/strategies-initiatives-and-awards/strategies-plans-and-visions/environment-and-energy/the-uae-net-zero-2050-strategy. Last accessed 28 Jul 2025
44. World Health Organization (2025) World report on ageing and health. https://iris.who.int/handle/10665/186463. Last accessed 28 Jul 2025
45. IPCC (2025) Sixth assessment report: regional fact sheet—Middle East https://www.ipcc.ch/report/ar6/wg1/chapter/chapter-12/. Last accessed 28 Jul 2025

Smart Policy for Inclusive Tech-Enabled Cities

Vikas Chand Sharma and Kirti Babber

Abstract The accelerated growth of technology in cities has brought with it unprecedented opportunities for innovation and economic progress, but at the same time, it has also heightened already existing disparities and produced new kinds of digital exclusion. The role of smart policy frameworks in guaranteeing that tech-driven urban development benefits all citizens equally is explored in this chapter. By reflection on present policy strategies, review of effective case studies, and integration of theoretical literature, we contend that inclusive tech-enabled urbanism calls for intentional policy action addressing access, participation, and benefit sharing. The chapter presents a holistic framework for smart policies that enhance equitable urban entrepreneurship while harnessing technology for inclusive city growth. Key conclusions indicate that effective inclusive tech policies need to combine community participation, support targeting mechanisms, and flexible governance arrangements in order to close digital divides and provide widespread urban prosperity.

Keywords Inclusive urbanism · Smart cities · Digital equity · Urban entrepreneurship · Technology policy · Urban governance

1 Introduction

The modern urban environment is increasingly being defined by the integration of digital technologies into the texture of urban life, making a core shift in how cities operate, how citizens engage with urban services, and how economic opportunities are allocated in urban space [1]. This phenomenon, which is more familiarly

V. C. Sharma (✉)
University Institute of Architecture, Chandigarh University, Mohali 140413, Punjab, India
e-mail: ar.vikassharma@gmail.com

K. Babber
University Centre of Research and Development, Chandigarh University, Mohali 140413, Punjab, India

V. Chand Sharma et al. (eds.), *Tech-Enabled Urbanism and Entrepreneurship for Inclusive Cities*, Lecture Notes in Civil Engineering 779,
https://doi.org/10.1007/978-981-95-3873-7_13

described as tech-enabled urbanism, includes a wide range of technological interventions, from smart infrastructure for cities and digital platforms of government to the spread of technology-enabled entrepreneurship and innovation environments [2]. Although these trends hold significant potential for making cities more efficient, sustainable, and of better quality, they also pose fundamental questions regarding equity, inclusion, and the sharing of benefits and costs of urban technological change.

The growth of technology-enabled urbanism has been spurred by several converging drivers, including swift advancements in information and communication technologies, accelerating rates of urbanization, rising urban sustainability and resilience concerns, and the advent of new economic models predicated on digital platforms and data-driven services [3]. Cities across the globe have adopted multiple types of smart city projects, aiming to utilize technology to overcome complicated urban challenges from traffic and energy efficiency to service provision and civic participation [4]. At the same time, cities are also centers of technology entrepreneurship where cities fight to host and cultivate clusters of innovation, startup environments, and digital economy activity [5]. But the embedding of technology in cities has not been equally advantageous for all population groups and geographic locations within cities. Studies have repeatedly confirmed the creation of digital divides that reproduce and even enlarge pre-existing socioeconomic disparities, producing new exclusion processes and processes of marginalization [6]. Poor communities, racial and ethnic minorities, older persons, and other marginalized groups frequently experience obstacles to their participation in and benefit from tech-enabled urban services and opportunities [7]. Moreover, the planning and execution of smart city projects tend to give priority to the interests of technology proponents and urban elites, often sidelining the needs and preferences of marginalized groups [8]. The challenge of fostering inclusivity in tech-enabled urbanism is the result of the continuous increase in the pace of technological change, the complex relationship between technology and urban social systems, and the socio-technical processes within them. Because of the dynamic and multifunctional characteristics of technology-saturated urban change, traditional urban planning and policy frameworks built for more settled and predictable cities are apt to struggle [9]. This, in turn, contributes to policy failure, emerging problems, and missed chances for using technology in fostering an inclusive urban development. Addressing these challenges has sparked greater focus on the development of smart policy frameworks intended to guide urban tech development in more inclusive and equitable directions. These policies will need to resolve the paradox of stimulating innovation and entrepreneurship while ensuring broad-based technological benefits to the urban population [10]. They have to address important governance issues such as protecting privacy, handling data ethically, encouraging participatory governance, and ensuring responsible civic tech in the urban governance ecosystem [11]. This chapter aims to analyze the strategic policy design that fosters inclusivity in urbanism at a tech-enabled level. It looks at existing policy gaps, aims to learn from policy successes, and makes policy design suggestions for the intended gaps. To structure the discussion, the chapter put forwards three primary research questions: (1) What design elements which have the most importance in constituting effective policies for inclusive tech-enabled urbanism? (2) How can policy structures create opportunities for fair and equitable urban entrepreneurial activities and at the same time leverage technology for higher societal developmental goals? (3) What policy lessons can be learned from current policy experience to inform future policy development in this area? The chapter is designed to offer a complete analysis of such questions using investigation of theoretical frameworks, policy analysis, case study research, and synthesis of recommendations.

2 Context

2.1 Tech-Enabled Urbanism

Tech-enabled urbanism is a paradigmatic change in the way cities function, evolve, and interact with their citizens through the conscious embedding of digital technologies into urban systems and processes [12]. It covers various aspects of urban experience, ranging from infrastructure and service provision to economic growth and social interaction. Technologically enabled urbanism is essentially the application of information and communication technologies to optimize urban functionality, efficiency, and quality of life and produce opportunities for innovation and economic growth [13].

The expressions of tech-powered urbanism are varied and changing fast. Smart city projects are one dominant manifestation, which entails embedding sensors, data analysis, and automated networks within urban infrastructure to streamline traffic, energy use, waste disposal, and other municipal services [14]. Digital platforms of governance are another critical aspect, which allows for online service provision, civic participation, and collaborative decision-making mechanisms [15]. The expansion of digital platforms, mobile applications, and location-based services has also reshaped the manner in which inhabitants move through and engage with urban spaces [16].

Most importantly, tech-enabled urbanism has come to be deeply linked with the growth of innovation districts, technology clusters, and entrepreneurial ecosystems that make cities the hubs of digital economy activity [17]. Urban policymakers have actively pursued these developments in efforts to attract investment, talent, and economic growth through technology-centered development strategies. The emergent urban environment is thus defined by the juxtaposition of traditional urban activities with novel forms of digitally mediated economic activity, social interactions, and service provision.

2.2 Inclusivity in Urban Contexts

Urban inclusivity is the degree to which every citizen, irrespective of socioeconomic status, demographics, or locality within the city, possesses substantial access to opportunities, services, and decision-making within cities [18]. Urban inclusivity involves several aspects of urban living, such as economic inclusion, social inclusion, spatial inclusion, and political inclusion. Economic inclusion means making sure that all citizens are able to access job opportunities, entrepreneurial support, and economic mobility paths. Social inclusion includes access to social services, educational opportunities, cultural experiences, and social networks that promote well-being at both the individual and community levels [19].

Spatial inclusion concerns the city-wide distribution of opportunities and resources, the understanding that where individuals reside affects their access to jobs, services, and amenities [20]. Political inclusion entails guaranteeing all citizens have real chances to engage in decision-making that impacts their communities and lives. These forms of inclusion are interdependent and reinforcing, exclusion in one often causing exclusion in the others.

The test of fostering inclusiveness in urban settings has grown more sophisticated in the age of globalization, economic restructuring, and technological transformation. Urban areas have become more economically polarized, and differences between high-skilled, high-paid workers and low-skilled, low-paid workers have widened [21]. Residential income, racial, and ethnic segregation has continued or has reinforced itself in many cities, and there are spatial concentrations of advantage and disadvantage. These trends have contributed to making it more challenging to promote equitable benefits from urban growth for all residents.

2.3 Intersection of Tech, Urbanism, and Inclusivity

Where technology intersects urbanism and inclusivity, there are both opportunities and challenges for facilitating urban development for all. On the one hand, technology can democratize access to information, services, and opportunities, facilitating more inclusive urban development [22]. Digital channels can offer new avenues for civic participation, enabling previously marginalized groups to engage in urban governance and decision-making. Technology-enabled services have the potential to enhance access to education, healthcare, financial services, and other social and economic mobility-enabling resources.

Technology can also facilitate more responsive and effective delivery of services, potentially enhancing outcomes for low-income and underserved populations. For instance, mobile health apps can bring healthcare to distant or underserved locations, while online education platforms can make learning available independent of geographic location or socioeconomic circumstance [23]. Likewise, technology-enabled entrepreneurship can establish new avenues to economic opportunity for those who are hindered from access to conventional employment or business development.

But technology integration also introduces new inequalities and exclusion in urban areas. The digital divide, with unequal access to digital skills and technology, serves to perpetuate and exaggerate existing socioeconomic inequalities [24]. Neighborhoods lacking access to broadband internet, digital equipment, or digital literacy may be barred from tech-enabled services and benefits. Furthermore, technological systems' design and implementation tend to be influenced by the biases and agenda of their inventors, often at the expense of excluding some groups or communities.

3 Policy Approaches

3.1 Review of Existing Smart City Policies

Current policies of smart cities demonstrate a broad range of approaches, priorities, and realization strategies, influenced by indigenous contexts, political ideologies, and accessible resources [25]. Most policies in place have similar goals of using technology to enhance urban efficiency, sustainability, and quality of life but differ widely in regard to equity and inclusion issues. Most early smart city projects emphasized technological rollout and system calibration with minimal regard for social effects or distributive consequences [26]. Smart city policies by government usually focus on infrastructure development, data integration, and digitalization of services. For instance, Singapore's Smart Nation initiative is an all-encompassing urban digitization strategy with investments in digital infrastructure, data analytics capacity, and digital services for citizens [27]. Just like Barcelona smart city policy has centered on utilizing technology to enhance urban sustainability, civic engagement, and service delivery with continued focus on social inclusion and democratic participation [28]. The application of public-private partnerships has increased in scope in relation to the deployment of smart city policies and framework systems, as they require substantial funding and expertise. As mentioned by "Kitchin and others, "technology firms are involved in the partnership by providing infrastructure and services in exchange for urban data and access to markets." These partnerships, while enabling more rapid adoption of technology and reducing public investment, raise concerns over corporate governance of city administrations and the extent to which private interests would dominate public objectives. Tech-enabled urban policies are focused on fostering environments conducive to building technology entrepreneurship and the digital economy. This is generally aimed at providing, as policy framework, innovation districts, incubators and accelerators, regulatory sandboxes, talent magnets, and innovation and talent magnets [29]. Austin, Boston, and Tel Aviv are examples of cities that have built well-rounded innovation ecosystems through focused policy interventions geared toward attracting and developing technology entrepreneurship.

3.2 *Policy Failures and Unintended Consequences*

In spite of popularizing smart city and tech-centric urban policies, various instances of policy failure and collateral damages underscore the difficulties involved with technology-driven urban development. Most smart city projects have either been unable to achieve expected dividends or generated new issues while trying to address the existing ones [30]. Such failures tend to arise due to excessively technocratic responses that emphasize technological solutions without proper comprehension of social and urban intricacies.

A typical failure pattern is where technology systems are introduced that fail to meet user requirements or local contexts. For instance, some cities have heavily invested in smart city systems which were not fully utilized because they were overly complicated for residents to utilize or did not respond to priorities recognized by community members [31]. Such a failure emphasizes the need for user-focused design and community input in the formulation of technology policy.

Problems of digital divide have arisen as an unforeseen impact of a lot of technology-enabled urban policies. While these policies tend to enhance services and opportunities access, they tend to create new barriers for residents who do not have access to appropriate technology or digital know-how [32]. For example, government services digitization can find it harder for older citizens, the poor, or others with minimal digital competence to access fundamental services.

Another category of unintended consequences associated with technology-oriented urban development policies is gentrification and displacement. The creation of tech clusters and innovation districts may raise living expenses and real estate values, which may force out existing businesses and residents [33]. Cities like San Francisco, where the growth of the technology sector has resulted in the widespread displacement of low-income households and communities of color, have seen this trend.

Key Challenge for Policy Makers

Addressing inequities and fostering innovation simultaneously is one quote "Mazzucato challenge of the policy-makers" [34]. Hurdle includes the goal of region-diagramming the effect of the innovation. One of the policy issues to be solved is the promotion of entrepreneurship within a region, and innovation should balance the framework so that the growth elements do not pose a negative effect to the socially responsible and sustainable developmental goals. Another issue that is important is the advancement of the new policy framework of the urbanism as to where new jurisdictions, sectors, and stakeholders are involved. Community and participatory governance is important to address the problems of the barrios and these issues as described need many government layers of participation [35]. The sudden and still unpredictable events caused within the society are also more than capable of creating hurdles to the previously accepted policy framework for the society, as these crisis or society's innovation surge require a leap away after for society's existing systems of planning and executing policies. Policymakers who want to use technology to support urban growth while protecting the rights and interests of residents face additional challenges due to data privacy and governance concerns. Urban data collection, storage, and use bring up complex privacy, consent, transparency, and accountability issues that require careful policy consideration. Legislators must develop frameworks that allow for constructive uses of urban data while offering appropriate safeguards and supervision. Lack of resources and capacity are practical challenges that many cities face when developing and implementing inclusive technology-enabled urban policies. It can be difficult for smaller and less wealthy cities to compete with larger, wealthier cities for talent and technology investment [36]. Such disparities could limit the potential broad-based benefit of technology-enabled urban development and exacerbate regional disparities.

4 Discussion

4.1 Smart and Inclusive Policy Frameworks

In order to address the unique opportunities and challenges of integrating technology into cities, traditional urban policy strategies must be critically rethought in order to create inclusive and intelligent policy systems for tech-enabled urbanism. These systems must have explicit commitments to equity and inclusion, be principled yet flexible, and be able to react quickly to changes in technology [37]. Since exclusive sectoral approaches cannot guarantee inclusive tech-enabled urbanism, effective frameworks integrate the various policy areas, including social policy, economic development, urban planning, and technology policy. Prioritizing community values and needs over technological potential is one of the fundamental tenets of intelligent and inclusive policy frameworks. Such frameworks identify community priorities and urban challenges first, then consider how technology can help address them, as opposed to starting with current technologies and searching for applications. Strong community engagement procedures that can record a range of opinions are necessary for this, as is the creation of policies based on the actual experiences of urban residents from various demographic groups and geographical areas.

A key element of inclusive policy frameworks are participatory governance mechanisms, which allow for continuous community input into the formulation and application of policies [38]. Community-based research projects, digital engagement platforms, participatory budgeting procedures, and citizen advisory committees are a few examples of these mechanisms. For residents with different levels of technological access and comfort to have access to participation opportunities, it is imperative that digital and non-digital engagement channels be integrated. Equity impact assessment tools add another critical component to smart and inclusive policy design, allowing policymakers to critically consider the probable distributional effects of proposed policies and interventions [39]. Assessments must take into account impacts on various facets of equity, such as access to opportunities, benefit and burden distribution, and impacts on various demographic groups and places. Routine monitoring and evaluation systems must be in place to monitor policy impacts and make informed adjustments to enhance equity performance.

Frameworks for cross-sector coordination are essential for addressing the complex governance conundrums associated with inclusive tech-enabled urbanism. These frameworks must guarantee clear lines of accountability and collaborative efforts among government sectors, agencies, and levels [40]. Establishing new institutional structures, such as innovation offices, digital equity coordinators, or cross-departmental task forces on inclusive technology policy, is often necessary for effective coordination.

Another crucial component of effective frameworks is adaptive design for policy, which recognizes that urban and technological environments are constantly changing [41]. Adaptive policies include cyclical learning mechanisms, experimental pilot programs, and tools for regular review and modification. This has the effect of enabling policymakers to develop evidence about what works in different environments while also adapting to changing circumstances.

4.2 *Equitable Urban Entrepreneurship*

Enabling equitable urban entrepreneurship entails policy strategies that overcome the structural barriers that inhibit certain groups and communities from engaging in entrepreneurial activities while leveraging current assets and capabilities within low-income communities. Conventional programs for supporting entrepreneurship tend not to reach or serve efficiently entrepreneurs in low-income communities, ethnic and racial minority communities, women, immigrants, and other minority groups [42]. Closing these gaps involves focused interventions that acknowledge and address the unique challenges experienced by various entrepreneurial groups.

Access to capital is one of the strongest obstacles to inclusive entrepreneurship, given that conventional funds tend to exclude entrepreneurs who do not have collateral, creditworthiness, or established business connections [43]. The establishment of alternative funding sources, such as public venture capital funds that target underrepresented entrepreneurs, microloan programs, crowdfunding websites, and community development financial institutions, may be one way for policymakers to address this issue. Oakland and Boston, among others, have established innovative financing programs that pool public and private funds to fund diverse entrepreneurs.

Business development and technical assistance services are also critical in fostering equal entrepreneurship, especially among entrepreneurs who lack past business experience or professional networks. Successful programs ensure culturally sensitive services in a variety of languages and formats, both online and in-person [44]. Mentorship initiatives offering pairings between novice entrepreneurs and veteran business leaders can offer advice while establishing social capital and professional connections.

Physical infrastructure and space requirements are yet another significant factor that should be addressed under equitable urban entrepreneurship policies. Most underrepresented entrepreneurs have limited access to low-cost commercial space or suitable facilities for business growth. Policies can respond to these requirements by establishing affordable commercial areas, shared-use facilities, equity-oriented business incubators, and zoning changes that facilitate small-scale entrepreneurial activities in residential and mixed-use zones [45].

Market access and procurement policies have significant potential to enable supportive equitable entrepreneurship through stimulating demand for products and services offered by underrepresented entrepreneurs. Supplier diversity programs, social procurement policies, and local business preference programs can be adopted by cities to promote purchases from minority-owned and local businesses [46]. Digital platforms and marketplaces can facilitate linking local entrepreneurs with customers and increasing market opportunities.

Access to technology and digital inclusion are becoming increasingly critical determinants of entrepreneurial success, as most business tasks and market opportunities are now digitally mediated. Policies must confront digital divides by providing access to broadband internet, digital equipment, and technology training programs [47]. Public Wi-Fi networks, computer labs, and digital literacy programs can serve as critical support for entrepreneurs who have no personal technology equipment.

Considerations in the regulatory environment are important in enabling fair entrepreneurship since cumbersome or complicated regulations can pose a hindrance to small and new firms. Regulations should ease business registration procedures, issue clear guidelines on regulative needs, and take note of the varying effects of regulations across varying firms and sizes [48]. Pilot programs and regulatory sandboxes may allow entrepreneurs to pilot innovative business models while collaborating with regulators to come up with suitable oversight methods.

Linking with more comprehensive economic development plans is necessary for making sure equitable entrepreneurship policies drive inclusive city growth instead of acting in a vacuum. The linkage must account for how entrepreneurship building relates to labor development, affordable housing, transportation, and other city systems that influence entrepreneurial success and community impacts [49]. Place-based strategies that concentrate entrepreneurship assistance in certain neighborhoods or communities can assist with making sure that local residents benefit from economic development gains.

Measurement and assessment systems are needed to monitor the effectiveness of equitable entrepreneurship policy and make adjustments as needed to enhance outcomes. Indicators should account for business outcomes (e.g., new business formation, survival rates, and revenue growth) as well as equity outcomes (e.g., demographic group and geographic area participation rates). Periodic assessment should guide policy refinement and assist in identifying model practices that can be scaled or replicated.

5 Case Studies

5.1 Study 1: Barcelona's Decidim Platform and Digital Sovereignty Initiative

Barcelona provides a compelling example of how cities can use technology for democratic engagement and participation while staying focused on equity and inclusion issues through its inclusive tech-enabled urbanism strategy, which is exemplified by the Decidim digital participation platform and the larger digital sovereignty project. Under the leadership of Mayor Ada Colau and the Barcelona En Comú coalition, the city has developed new digital governance models that prioritize openness, control over technology, and participatory citizen engagement. Launched in 2016, the Decidim platform is an open-source digital framework for participatory democracy that makes it easier for citizens to participate in various forms of democratic engagement, including collective policy-making, policy consultation, citizen assemblies, and participatory budgeting [50]. Multilingualism, accessibility features, and integration with offline participation mechanisms were all included in the platform's development to ensure that a wide range of groups could take advantage of digital participation opportunities. To ensure the Decidim platform would accommodate a range of user needs and preferences, residents, community organizations, tech developers, and city employees were brought together for extensive community participation and co-design processes. To help residents use the platform, the city invested a lot of money in community outreach, digital literacy programs, and capacity building. These initiatives were especially designed to reach out to long-standing marginalized groups, including low-income communities, immigrants, and senior citizens.

Barcelona's approach has produced remarkable results in terms of policy changes and participation rates. The Decidim platform has over 400,000 registered citizens, and participation in policy surveys and participatory budgeting exercises is particularly high [51]. The platform has been used to develop important city policies, including the city's climate action plan and the traffic management superblocks program. Notably, analysis of the participation patterns shows comparatively broad participation across different neighborhoods and demographic groups, suggesting some success with inclusive participation. Initiatives to reduce reliance on proprietary technology platforms, create open-source alternatives, and maintain democratic control over urban technology systems have all been part of the broader digital sovereignty project. This has included funding community broadband networks, supporting platform cooperatives, and, when necessary, implementing open-source procurement practices [52]. These initiatives aim to address power imbalances in urban technological systems and stop technological advancements from serving the interests of the community rather than corporate profits. Among the most significant takeaways from the Barcelona experience are the following: care to ensure accessibility and inclusion in platform design; high levels of investment in community participation and capacity building; political leadership with a stake in participatory government; and connecting digital participation to democratic governance processes more generally. The Barcelona experience demonstrates that, when used responsibly for equity and community ownership, technology can be used to enhance rather than replace established democratic processes.

5.2 *Case Study 2: Medellin's Innovation District and Social Urbanism*

The transformation of Medellín from a violent and unstable city to a leader in urban innovation worldwide offers important insights into how entrepreneurship and technology can be used to create inclusive cities. In order to address inequality and promote social inclusion, the city has implemented a strategy known as "social urbanism," which combines infrastructure investment, educational interventions, and the creation of innovation districts [53].

Situated in the Ruta N district, the Medellín Innovation District is an integrated approach to technology-driven economic development that aims to connect innovation endeavors with broader urban development. The district has been planned as part of a larger city transformation strategy that includes investments in public transportation, education, cultural facilities, and public space, as opposed to developing a standalone technology park. [54]. This all-encompassing approach takes into account that cities that provide a high standard of living for a variety of groups are more conducive to innovation and entrepreneurship.

Medellín's strategy has placed a strong emphasis on education and workforce development, investing heavily in educational institutions, programs, and facilities. In low-income neighborhoods, the city has built state-of-the-art educational facilities like library parks that combine libraries, schools, and community centers [55]. Workforce training initiatives that prepare locals for positions in the growing technology sector have been added to these.

Instead of just attracting outside investment and talent, Medellín's entrepreneurship support programs have concentrated on fostering local capacity and community-based innovation. Local entrepreneurs have benefited from initiatives like Ruta N's entrepreneurship activities, which have connected them to global markets and networks [56]. Additionally, the city made investments in innovation labs, maker spaces, and other community-based infrastructure that supports entrepreneurship and innovation.

Medellín's efforts have paid off, as evidenced by notable improvements in quality-of-life indices, innovation levels, and economic development metrics. The city has established a burgeoning startup scene and drawn significant technology investment. Notably, social advancements like improved educational attainment, decreased crime, and less inequality have coincided with these economic development advancements [57]. However, there are still issues with ensuring that the benefits of innovation district development are available to all communities and citizens of cities.

Concerns about the long-term sustainability of the city's development model persist, and some argue that the expansion of the technology sector has pushed gentrification in some places [58]. These problems highlight the ongoing need for innovation-led urban development to prioritize equity and inclusion. A few key takeaways from Medellín's experience include the necessity of integrating entrepreneurship and innovation development with broader urban development strategies, investing heavily in education and human capital development, emphasizing place-based assets and capabilities, and having a strong political will committed to inclusive development. The city's approach shows that, when implemented as part of integrated strategies that address multiple facets of urban development simultaneously, technology-oriented economic development can be a component of inclusive urban development.

6 Recommendations

The analysis of existing policy frameworks, examination of case studies from Barcelona and Medellín, and identification of persistent implementation challenges reveal the need for actionable policy interventions that move beyond theoretical constructs toward practical implementation. Current smart city initiatives often fail not due to inadequate technology, but because of insufficient attention to governance structures, community engagement mechanisms, and equity-focused design principles. The policy failures documented across multiple urban contexts demonstrate that technocratic approaches without deliberate inclusion strategies consistently reproduce existing inequalities while creating new forms of digital exclusion.

Drawing from the successful elements of Barcelona's participatory governance model and Medellín's integrated social urbanism approach, effective policy design must address five critical domains simultaneously rather than treating them as separate initiatives. Cities cannot achieve inclusive tech-enabled urbanism through fragmented approaches that address digital access without considering entrepreneurship opportunities, or that promote innovation districts without ensuring community participation in their development. The interconnected nature of urban technological change requires coordinated policy responses that recognize how governance decisions, community engagement processes, access barriers, economic development strategies, and evaluation systems influence each other.

The recommendations presented here emerge from the gap between current policy practices and the requirements for equitable technological urban development. They reflect lessons learned from policy implementations that have successfully navigated the tension between promoting innovation and ensuring inclusive benefits distribution. These interventions are designed to be implementable within existing municipal governance structures while providing the flexibility needed to adapt to local contexts and changing technological landscapes.

Rather than proposing entirely new institutional frameworks, these recommendations focus on modifying existing policy processes and creating new coordination mechanisms that can be integrated into current urban governance systems. Each

Table 1 Policy recommendations for inclusive tech-enabled urbanism

Policy area	Actions	Implementation tools	Metrics
Digital governance	Digital equity offices, Cross-departmental task forces, Equity impact assessments	Standard assessment frameworks, Regular coordination meetings, Community partnership agreements	Policy coherence scores, Cross-sector collaboration index, Community representation rates
Community participation	Multi-channel engagement, Digital literacy programs, Resident-led research funding	Online/offline participation platforms, Neighborhood-based training hubs, Participatory budgeting tools	Participation rates by demographics, Digital skills improvement, Policy influence tracking
Access and inclusion	Affordable broadband programs, Multilingual service delivery, Accessibility compliance	ISP partnership agreements, Community translation networks, Assistive technology integration	Digital divide reduction, Service accessibility scores, Usage rates by language groups
Local entrepreneurship	Alternative financing mechanisms, Procurement preferences, Business incubation spaces	Microloan funds, Supplier diversity quotas, Shared commercial facilities	Business formation rates, Funding access by demographics, Local procurement percentages
Performance monitoring	Real-time equity dashboards, Community evaluation panels, Adaptive policy cycles	Data collection systems, Resident advisory committees, Quarterly review processes	Equity outcome trends, Community satisfaction scores, Policy adjustment frequency

Source By Authors

recommendation includes specific implementation tools and measurable outcomes to enable cities to assess progress and make necessary adjustments based on local conditions and community feedback. The emphasis on concrete actions and measurable metrics reflects the need for policy approaches that can demonstrate tangible improvements in equity outcomes while maintaining the innovation potential that makes tech-enabled urbanism attractive to municipal leaders and community members alike (Table 1).

7 Conclusion

Given the rapid advancement of technology and persistent urban inequality, this chapter has examined the critical role that well-thought-out policy designs play in promoting inclusive tech-facilitated urbanism. The conversation demonstrates how technology can significantly improve living standards, sustainability, and urban efficiency, but achieving these benefits equitably calls for deliberate policy intervention that is in line with access, participation, and benefit distribution among diverse urban groups. The majority of smart city and tech-oriented urban policies fail to fully take equity and inclusion issues into account, according to an analysis of current policy strategies, which reveals significant gaps in current frameworks. The significance of adopting a more comprehensive and introspective approach when designing technology policies is highlighted by policy flaws and unintended consequences like digital divides, gentrification forces, and the exclusion of marginalized groups.

Cities can use technology to promote democratic participation and equitable economic growth if policies are designed with particular attention to equity and community involvement, as demonstrated by the case studies of Barcelona and Medellín. Several significant guiding principles for policy development in this area are produced by the analysis of clever and inclusive policy frameworks. They include putting the needs of the community before the capabilities of technology, establishing robust procedures for community involvement, developing systems of adaptive governance, and maintaining a methodical focus on the effects of equity as policies are implemented. The conversation about equitable urban entrepreneurship highlights the need for targeted interventions that take advantage of the resources and skills already available in the community while also removing systemic obstacles.

This chapter's suggestions provide a comprehensive manual for cities looking to develop intelligent policies for inclusive tech-enabled urbanism. The recommendations emphasize the value of cross-sector coordination-capable institutional arrangements, efficient community engagement procedures, integrated digital inclusion frameworks, the promotion of equitable entrepreneurship, and reliable monitoring and assessment systems. It will take a significant amount of political will, resource allocation, and ongoing attention to equity outcomes to put these recommendations into practice. Future research will need to concentrate on a number of themes in order to advance our understanding of inclusive tech-enabled urbanism. First, in order to assess the long-term impacts of different policy approaches on equity outcomes, community formation, and urban sustainability, longitudinal analysis is crucial.

In addition to tracking changes in digital divides and other inclusion metrics over time, longitudinal studies should examine the intended and unintended consequences of technology policy.

Second, comparing different urban contexts would help determine how local factors such as political cultures, economic conditions, demographics, and governance structures affect the effectiveness of inclusive technology policies. In order to determine how these larger institutional and policy environments influence local policy options and outcomes, such studies ought to incorporate cities from a range of national contexts. Third, more research is required on the governance concerns surrounding the regulation of technological change in urban areas, including concerns about privacy protection, democratic accountability, and community management of technology systems. This study would investigate novel governance models and assess how they affect democratic participation and technological advancement.

The relationship between technology policy and other urban policy areas, such as housing, transportation, education, and environmental policy, also requires further study. In order to create integrated approaches to urban development that can handle several issues at once and prevent unforeseen consequences across policy areas, it is imperative to comprehend these intersections. According to this chapter's findings, inclusive tech-enabled urbanism is both feasible and necessary, but it requires significant changes in how cities approach technology policymaking. For urban technological change to fairly benefit all urban dwellers and support more equitable and sustainable urban futures, steps must be taken to move beyond technocratic approaches and toward more participatory, equity-oriented, and community-led policy styles.

References

1. Kitchin R (2014) The data revolution: big data, open data, data infrastructures and their consequences. Sage
2. Townsend AM (2013) Smart cities: Big data, civic hackers, and the quest for a new Utopia. WW Norton and Company
3. Batty M (2013) The new science of cities. MIT Press
4. Albino V, Berardi U, Dangelico RM (2015) Smart cities: definitions, dimensions, performance, and initiatives. J Urban Technol 22(1):3–21. https://doi.org/10.1080/10630732.2014.942092
5. Florida R (2017) The new urban crisis: how our cities are increasing inequality, deepening segregation, and failing the middle class-and what we can do about it. Basic Books
6. Robinson L, Cotten SR, Ono H, Quan-Haase A, Mesch G, Chen W, Stern MJ (2015) Digital inequalities and why they matter. Inf Commun Soc 18(5):569–582. https://doi.org/10.1080/1369118X.2015.1012532
7. Helsper EJ (2012) A corresponding fields model for the links between social and digital exclusion. Commun Theory 22(4):403–426. https://doi.org/10.1111/j.1468-2885.2012.01416.x
8. Hollands RG (2020) Will the real smart city please stand up? In: Willis KS, Aurigi A (eds) The Routledge companion to smart cities. Routledge, pp 179–199
9. Foth M, Choi JH, Satchell C (2011) Urban informatics. In: Proceedings of the ACM 2011 conference on computer supported cooperative work. Association for computing machinery, pp 1–8. https://doi.org/10.1145/1958824.1958826
10. Cardullo P, Kitchin R (2019) Being a 'citizen' in the smart city: up and down the scaffold of smart citizen participation in Dublin, Ireland. GeoJournal 84:1–13. https://doi.org/10.1007/s10708-018-9845-8
11. Goodman EP, Powles J (2019) Urbanism under google: lessons from Sidewalk Toronto. Fordham Law Rev 88:457
12. Shelton T, Zook M, Wiig A (2015) The 'actually existing smart city.' Camb J Reg Econ Soc 8(1):13–25. https://doi.org/10.1093/cjres/rsu026
13. Caragliu A, Del Bo C, Nijkamp P (2011) Smart cities in Europe. J Urban Technol 18(2):65–82. https://doi.org/10.1080/10630732.2011.601117
14. Nam T, Pardo TA (2011) Conceptualizing smart city with dimensions of technology, people, and institutions. In: Proceedings of the 12th annual international digital government research conference: digital government innovation in challenging times. Association for computing machinery, pp 282–291. https://doi.org/10.1145/2037556.2037602
15. Janssen M, Estevez E (2013) Lean government and platform-based governance—doing more with less. Gov Inf Q 30(Suppl. 1):S1–S8. https://doi.org/10.1016/j.giq.2012.11.003

16. de Waal M (2015) The future of the city: a smart city or a social city? Art Identity Polit 12:11–30. https://revistas.um.es/reapi/article/view/236041
17. Katz B, Wagner J (2014) The rise of innovation districts: a new geography of innovation in America. Brookings Institution
18. Fainstein SS (2010) The just city. Cornell University Press
19. Silver H (2015) The contexts of social inclusion (UN Document ST/ESA/2015/DWP/144). United Nations Department of Economic and Social Affairs. https://www.un.org/development/desa/dspd/2015/10/the-contexts-of-social-inclusion/
20. Soja EW (2010) Seeking spatial justice. University of Minnesota Press
21. Sassen S (2001) The global city: New York, London. Princeton University Press, Tokyo
22. Gurstein M (2003) Effective use: a community informatics strategy beyond the digital divide. First Monday 8(12). https://doi.org/10.5210/fm.v8i12.1107
23. Toyama K (2011) Technology as amplifier in international development. In: Proceedings of the 2011 iConference. Association for computing machinery, pp 75–82. https://doi.org/10.1145/1940761.1940772
24. Norris P (2003) Digital Divide: civic engagement, information poverty, and the internet worldwide. Can J Commun 28(1). https://doi.org/10.22230/cjc.2003v28n1a1352
25. Anthopoulos L (2017) Smart utopia vs smart reality: learning by experience from 10 smart city cases. Cities 63:128–148. https://doi.org/10.1016/j.cities.2016.10.005
26. Wiig A (2015) IBM's smart city as techno-utopian policy mobility. City 19(2–3):258–273. https://doi.org/10.1080/13604813.2015.1016275
27. Kong L, Woods O (2018) The ideological alignment of smart urbanism in Singapore: critical reflections on a political paradox. Urban Stud 55(4):679–701. https://doi.org/10.1177/0042098017746528
28. March H, Ribera-Fumaz R (2016) Smart contradictions: the politics of making Barcelona a self-sufficient city. Eur Urban Reg Stud 23(4):816–830. https://doi.org/10.1177/0969776414554488
29. Stam E (2015) Entrepreneurial ecosystems and regional policy: a sympathetic critique. Eur Plan Stud 23(9):1759–1769. https://doi.org/10.1080/09654313.2015.1061484
30. Luque-Ayala A, Marvin S (2015) Developing a critical understanding of smart urbanism? Urban Stud 52(12):2105–2116. https://doi.org/10.1177/0042098015577319
31. Vanolo A (2014) Smartmentality: the smart city as disciplinary strategy. Urban Stud 51(5):883–898. https://doi.org/10.1177/0042098013494427
32. Hargittai E (2001) Second-level digital divide: mapping differences in people's online skills (TPRC-2001-083). https://doi.org/10.48550/arXiv.cs/0109068
33. Zuk M, Bierbaum AH, Chapple K, Gorska K, Loukaitou-Sideris A (2018) Gentrification, displacement, and the role of public investment. J Plan Lit 33(1):31–44. https://doi.org/10.1177/0885412217716439
34. Mazzucato M (2018) The value of everything: making and taking in the global economy. Hachette UK
35. Pierre J (ed) (2011) The politics of urban governance. Palgrave Macmillan
36. Glaeser EL, Hausman N (2020) The spatial mismatch between innovation and joblessness. Innov Policy Econ 20(1):1–32. https://doi.org/10.1086/705646
37. Evans J, Karvonen A, Raven R (2016) The experimental city: new modes and prospects of urban transformation. In: The experimental city, 1st ed. Routledge
38. Fung A (2006) Varieties of participation in complex governance. Public Adm Rev 66(s1):66–75. https://doi.org/10.1111/j.1540-6210.2006.00667.x
39. Braveman PA, Kumanyika S, Fielding J, LaVeist T, Borrell LN, Manderscheid R, Rowe C (2011) Health disparities and health equity: the issue is justice. Am J Public Health 101(S1):S149–S155. https://doi.org/10.2105/AJPH.2010.300062
40. Jessop B (2002) The future of the capitalist state. Polity Press
41. Raven R, Schot J, Berkhout F (2012) Space and scale in socio-technical transitions. Environ Innov Soc Trans 4:63–78. https://doi.org/10.1016/j.eist.2012.08.001

42. Fairlie RW, Robb AM (2008) Race and entrepreneurial success: Black-, Asian-, and White-owned businesses in the United States. The MIT Press. https://doi.org/10.7551/mitpress/7961.001.0001
43. Blanchflower DG, Levine PB, Zimmerman DJ (2003) Discrimination in the small-business credit market. Rev Econ Stat 85(4):930–943. https://doi.org/10.1162/003465303772815835
44. Servon LJ (2017) The unbanking of America: how the new middle class survives. Houghton Mifflin Harcourt
45. Lerner J (2009) Boulevard of broken dreams: Why public efforts to boost entrepreneurship and venture capital have failed–and what to do about it. Princeton University Press. https://doi.org/10.1515/9781400831630
46. Ram M, Smallbone D (2003) Policies to support ethnic minority enterprise: the English experience. Entrep Reg Dev 15(2):151–166. https://doi.org/10.1080/0898562032000075177
47. Horrigan JB (2016) Digital readiness gaps. Pew research center
48. Dennis Jr WJ (2011) Entrepreneurship, small business and public policy levers. J Small Bus Manag 49(1):92–106. https://doi.org/10.1111/j.1540-627X.2010.00316.x
49. Porter M (2015) The competitive advantage of the inner city. In: LeGates RT, Stout F (eds) The city reader, 6th ed. Routledge, pp 314–327
50. Barandiaran XE, Calleja-López A, Monterde A, Romero C (2024) Decidim, a technopolitical network for participatory democracy: philosophy, practice and autonomy of a collective platform in the age of digital intelligence. Springer Nature Switzerland. https://doi.org/10.1007/978-3-031-50784-7
51. Peña-López I, Congosto M, Aragón P (2014) Spanish Indignados and the evolution of the 15M movement on Twitter: towards networked para-institutions. J Span Cult Stud 15(1–2):189–216. https://doi.org/10.1080/14636204.2014.931678
52. Morozov E, Bria F (2018) Rethinking the smart city: democratizing urban technology. Rosa Luxemburg Stiftung
53. Maclean K (2015) Social urbanism and the politics of violence: the Medellín miracle. Springer
54. Echeverri A, Orsini FM (2010) Informalidad y urbanismo social en Medellín. Environment, urbanism and society, Medellín, pp 130–152
55. Fuentes Villabona MB (2017) The role of the shared public realm in social equality: Medellin and 'social urbanism' [Master's thesis, McGill University]. eScholarship@McGill. https://escholarship.mcgill.ca/concern/theses/x633f848g?locale=en
56. Quinones G, Nicholson B, Heeks R (2015) A literature review of e-entrepreneurship in emerging economies: positioning research on Latin American digital startups. In: La Rovere RL, Ozório LDM, Melo LDJ (eds) Entrepreneurship in BRICS. Springer, pp 245–267. https://doi.org/10.1007/978-3-319-11412-5_11
57. Brand P (2013) Governing inequality in the South through the Barcelona model: 'Social urbanism. In: Medellín, Colombia. Interrogating Urban Crisis, pp 19–36
58. Behrendt F, Sheller M (2023) Mobility data justice. Mobilities 19(1):151–169. https://doi.org/10.1080/17450101.2023.2200148

MIX
Papier aus verantwortungsvollen Quellen
Paper from responsible sources
FSC® C105338

If you have any concerns about our products,
you can contact us on
ProductSafety@springernature.com

In case Publisher is established outside the EU,
the EU authorized representative is:
Springer Nature Customer Service Center GmbH
Europaplatz 3, 69115 Heidelberg, Germany

Printed by Libri Plureos GmbH
in Hamburg, Germany